A NATURALIST'S GUIDE

BUTTERFLIES
& DRAGONFLIES
OF
SRI LANKA

Gehan de Silva Wijeyeratne

JOHN BEAUFOY PUBLISHING

Reprinted in 2024

This edition published in the United Kingdom in 2018 by John Beaufoy Publishing Ltd
11 Blenheim Court, 316 Woodstock Road, Oxford OX2 7NS, England
www.johnbeaufoy.com

10 9 8 7 6 5 4 3 2

Photo captions and credits
Front cover: *Main image*: Tailed Jay *bottom left*: Common Tiger *bottom centre*: Red-veined Darter
bottom right: Tamil Lacewing. **Back cover:** Indian Sunbeam
Title page: Common Birdwing. **Contents page**: Rivulet Tiger © Matjaž Bedjanič
Main descriptions: photos are denoted by a page number followed by t (top), m (middle), b (bottom), l (left)
or r (right).
Matjaž Bedjanič 120b, 126tr, 134b, 136tl, 138tr, 138bl, 140t, 143t, 154bl, 166t **Karen Conniff** 118b, 125t,
126tl, 131, 132b, 136tr, 138tl, 143b, 150tr, 154tr, 159b, 163tl, 163b **Chitral Jayatilaka** 79b, 90b;
Ajith Ratnayaka 146bl, 152b, 155t, 155b, 158m, 162t, 163t; **George van der Poorten** 26b, 36t, 55b, 80t, 96t.

ISBN 978-1-912081-89-9

Edited by Krystyna Mayer
Designed by Gulmohur Press, New Delhi

Printed and bound in Malaysia by Times Offset (M) Sdn. Bhd.

·Contents·

INTRODUCTION

Increasingly, the segmentation between birders, butterfly watchers, dragonfly watchers and photographers is reducing as interests overlap and there is a demand for books that cover the three popular groups of birds, butterflies and dragonflies. Having already written the Sri Lankan bird title in this series, I was keen to produce a single, compact and portable photographic guide to the butterflies and dragonflies.

The 148 butterflies and 78 dragonflies described here represent just over 60 per cent of the butterfly and dragonfly species recorded in Sri Lanka. More importantly, with the emphasis on the most common species, the book covers around 90 per cent of the species that a visitor is likely to see. It also enables residents to learn about the most common species before progressing to more advanced technical books.

The book's two parts are presented as mini photographic field guides, each with its own introductory section. The focus is on field use for beginners and experts alike, with information on identifying each species and on its distribution and habitats. There is information on key sites, and the biology of dragonflies and butterflies, as well as up-to-date checklists with local status. Useful references for readers who wish to progress further in learning about these charismatic and photogenic animals are included at the end of the book. In brief, the book satisfies the following criteria:

• Compact and lightweight for field use.
• Identification oriented.
• Covers roughly 90 per cent of species likely to be encountered by casual butterfly and dragonfly watchers.
• Photographs show species the way they are seen in the field.

CLASSIFICATION OF SPECIES

Identifying species in a group, whether they be birds, mammals, butterflies or dragonflies, becomes easier when you are familiar with different 'types' of animal. The tree of life attempts to explain the interrelationship of species. For different animal groups, the levels used can vary and sometimes for the same group different authors have different views on the levels in the hierarchy. Controversy and debate about classification are many, and as a result different books classify species differently. Classification is also seemingly complicated by the use of suborders, superfamilies, families and subfamilies, within which different authors arrange species into different hierarchies. Butterflies are even more complicated, with the use of tribes. However, these are all ways of trying to map species into a tree of life to understand how they are related to each other through the act of evolution over time. Most people can ignore the classification hierarchies and be content with simply getting a feel of roughly where things belong in terms of simple relationships at family and order level.

The common name of a species is followed by its Latin name, which generally has two parts, the genus and the specific epithet. No two species can have the same Latin binomial, and the Latin names are relatively stable – a species should have only one accepted Latin name at any given time. Common names, on the other hand, can vary a lot from one country to another or even within regions in a country. A species may have a trinomial where it has been described as a subspecies or a geographical race. For example, the Purple-faced Leaf Monkey, an endemic mammal species in Sri Lanka, has four distinct subspecies. It logically follows that all four subspecies are endemic.

Meanwhile the Sri Lanka Cascader *Zygonix iris ceylonicus* is found only as a single subspecies, which is endemic. It has subspecies status to distinguish it from other populations found elsewhere in the world similar enough to be considered the same species, but different enough to be treated as a distinct subspecies.

Islands often have subspecies that are endemic as a result of isolation. Endemic subspecies on islands are of great interest to 'listers' as they are candidates for splits into new species. (By listers I mean wildlife enthusiasts like birders who want to maximize the number of species, and especially the number of endemic species, they have seen at home or abroad. Especially when travelling abroad, they will want to see an endemic subspecies in case it is later split into an endemic species.)

INSECTS

The Kingdom Animalia does not include single-cell animals and comprises phyla (singular phylum). Butterflies and moths are winged insects in the phylum Arthropoda. The major classes in this phylum include Crustacea (for instance crabs), Arachnida (spiders), Insecta (butterflies, beetles and other insects), and Myriapoda (centipedes, millipedes and similar).

The class Insecta is the largest of all the classes, with around a million species described. It is further divided into 29 orders that include the Odonata (dragonflies and damselflies), Lepidoptera (butterflies and moths), Orthoptera (grasshoppers), Coleoptera (beetles) and Hymonoptera (bees, wasps). To illustrate the levels of classification in the tree of life, the levels for butterflies and dragonflies are given below.

Domain	Eukaryota	Cellular organisms with a nucleus.
Kingdom	Animalia	Animals.
Subkingdom	Eumetazoa	All animals except a few groups such as sponges.
Phylum	Arthropoda	Invertebrate animals characterized by an outer skeleton, segmented body and jointed appendages.
Subphylum	Hexapoda	Insects and three groups of wingless arthropods.
Class	Insecta	All insects.
Subclass	Pterygota	Winged insects (or animals whose ancestors had wings).
Order	Lepidoptera	Butterflies and moths.
Suborder	Glossata (aka Rhopalocera)	**Butterflies.**
Suborder	Heterocera	Moths.
Order	Odonata	Dragonflies and damselflies.
Suborder	Zygoptera	**Damselflies.**
Suborder	Anisoptera	**True dragonflies.**

CLASSIFICATION OF BUTTERFLIES

The classification of butterflies remains in a state of flux as taxonomists revise relationships using new techniques in molecular phylogenetics. An example of a modern classification is shown below. The superfamilies Hesperoidea and Papilionoidea contain 33 subfamilies. Remarkably, of these 22 subfamilies are found in Sri Lanka. The subfamilies absent in Sri Lanka are shaded in grey.

Superfamily Hesperoidea (1 family, 6 subfamilies)	
Family Hesperiidae	
Subfamily Hesperiinae	Grass skippers or swifts
Subfamily Trapezitinae	Trapezitin skippers
Subfamily Heteropterinae	Skipperlings
Subfamily Pyrrhopyginae	Firetail skippers or firetips
Subfamily Pyrginae	Flats or spread-winged skippers
Subfamily Coeliadinae	Awls

Superfamily Papilionoidea (5 families, 27 subfamilies)	
Family Papilionidae	
Subfamily Parnassiinae	Snow apollos
Subfamily Papilioninae	Swallowtails
Subfamily Baroniinae	Baronia
Family Pieridae	
Subfamily Dismorphiinae	Mimic sulphurs
Subfamily Coliadinae	Sulphurs or yellows
Subfamily Pierinae	Whites
Family Lycaenidae	
Subfamily Theclinae	Hairstreaks, elfins and allies
Subfamily Lycaeninae	Coppers
Subfamily Polyommatinae	Blues

Subfamily Poritiinae	Gems
Subfamily Miletinae	Brownwings or harvesters
Subfamily Curetinae	Sunbeams
Subfamily Lipteninae	
Subfamily Liphyrinae	
Family Riodinidae	
Subfamily Euselasiinae	Neotropical metalmarks
Subfamily Riodininae	Metalmarks
Family Nymphalidae	
Subfamily Libytheinae	Beaks or snouts
Subfamily Danainae	Milkweed butterflies
Subfamily Morphinae	Morphos, owl butterflies and allies
Subfamily Satyrinae	Satyrids or browns
Subfamily Charaxinae	Leaf-wings
Subfamily Biblidinae	Tropical brushfoots and allies
Subfamily Apaturinae	Emperors
Subfamily Nymphalinae	True brushfoots
Subfamily Limenitidinae	Admirals and allies
Subfamily Acraeinae	Longwings
Subfamily Heliconiinae	Heliconians
Superfamily Hedyloidea (1 family)	
Family Hedylidae	American moth-butterflies

The above classification is at a relatively simple higher order. Subfamilies are further divided into tribes ending in 'ini'. For example, J. N. Eliot in his 1973 monograph divided the subfamily Polymatinnae into the four tribes Lycaenesthini, Candaladini, Niphandini and Polyomatini. The last tribe then had the genera allocated by him into 30 'sections'. More modern phylogenetic classifications are not that dissimilar and they continue to change.

Classification of Dragonflies

Dragonflies and damselflies belong to an order of animals known as the Odonota. This in turn comprises three suborders, the Zygoptera or damselfies, the Anisoptera or true dragonflies, and the Anisozygoptera, represented by a single family with one genus that has just two species. All odonates have two pairs of wings. In damselflies the front and rear wings are the same, and the term Zygoptera means equal wings. In dragonflies the front and rear wings are very different, and the term Anisoptera means unequal wings.

The venation of the wings in dragonflies is characteristic for different families. Specialists study the wing venation. Fortunately, most species have sufficient field characteristics to enable identification without resorting to examining the venation in the hand. This guide therefore ignores the wing venation.

The classification system below is based on data from Klaas-Dowuwe B. Dijkstra and others in a paper published in 2013. Because new dragonfly species are being described every year, the species and genera numbers are changing all the time. Here the numbers in the 2013 paper have been retained, as they provide an indication of the relative species richness of different superfamilies and families. In a further paper in 2014, Klaas-Dowuwe B. Dijkstra and others redefined the relationships of the damselflies. A number of changes were made which in a Sri Lankan context include the family Protoneurdae (the threadtails) being disbanded and moved into the subfamily Disparoneurinae, in the family Platystictidae. The genus *Onychargia* was moved into a new subfamily, Onychargriinae, and moved into the family Platystictidae. The species in the family Platystictidae found in Sri Lanka were moved into a new subfamily, Platystictinae, confined to Sri Lanka. A new subfamily, Protosticitinae, now contains the other species in the rest of the world. The authors have reinstated the genus *Ceylonosticta* for the genus *Drepanosticta*. The genera *Epophthalmia* and *Macromidia* in the family Corduliidae were moved into a new family, the Macromiidae, while the treatment of the genus *Macromidia* remains open.

The classification by Dijkstra and others has resulted in the order Odonata having three suborders, ten superfamilies and 30 families. This is summarized below, with the families not found in Sri Lanka being shaded in pale grey. For the reasons given above, the family Corduliidae is shaded in grey although in the species descriptions in this book three species are shown under the traditional arrangement of this family, rather than in the family Macromiidae. The paper in 2014 splits the damselflies into tribes – this is not shown here for simplicity.

Suborder Zygoptera (4 superfamilies, 18 families)		
Superfamily Lestoidea	Genera	Species
Family Hemiphlebiidae	1	1
Family Perilestidae (shortwings)	2	19
Family Synlestidae (giant spreadwings)	9	39
Family Lestidae (spreadwings)	9	151
Superfamily Platystictoidea	Genera	Species
Family Platystictidae (shadowdamsels)	6	224

Superfamily Calopterygoidea	Genera	Species
Family Amphipterygidae	4	14
Family Calopterygidae (demoiselles)	21	185
Family Chlorocyphidae (jewels)	19	144
Family Dicteriadidae	2	2
Family Euphaeidae (satinwings)	12	68
Family Lestoideidae (bluestreaks)	2	9
Family Megapodagrionidae (flatwings)	42	296
Family Philogangidae	1	4
Family Polythoridae	7	59
Family Pseudolestidae	1	1
Superfamily Coenagrionoidea	Genera	Species
Family Isostictidae	12	46
Family Platycnemididae (featherlegs)	40	404
Family Coenagrionidae (pond damselflies)	114	1,267
Suborder Anisozygoptera (1 Superfamily, 1 Family)		
Superfamily Epiophlebioidea	Genera	Species
Family Epiophlebiidae	1	2
Suborder Anisoptera (5 Superfamilies, 11 Families)		
Superfamily Aeshnoidea	Genera	Species
Family Austropetaliidae	4	11
Family Aeshnidae (hawkers)	51	456
Superfamily Petaluroidea	Genera	Species
Family Petaluridae (petaltails)	5	10
Superfamily Gomphoidea	Genera	Species
Family Gomphidae (clubtails)	87	980
Superfamily Cordulegastroidea	Genera	Species
Family Chlorogomphidae	3	47
Family Cordulegastridae (spiketails)	3	46
Family Neopetaliidae	1	1
Superfamily Libelluloidea	Genera	Species
Family Synthemistidae (tigertails)	9	46
Family Macromiidae (cruisers)	4	125
Family Corduliidae (emeralds)	20	154
Family Libellulidae (skimmers)	142	1,037
Classification uncertain	19	98

BUTTERFLIES

The distinction between butterflies and moths is somewhat artificial, although convenient. There are no absolute differences between butterflies and moths although there are some general ones, to which there are a few exceptions. Butterflies fly by day and moths fly by night. Butterflies are generally brightly coloured. There are, however, day-flying moths and some are colourful. The antenna of a butterfly has a club tip, whereas in moths the tip may be feathery or thick, and is hardly ever club tipped. Moths are further distinguished by having a coupling mechanism that links the forewings and hindwings in flight. Only one species of butterfly, an Australian skipper, has this.

STRUCTURE OF A BUTTERFLY

The body of an adult butterfly is broadly divided into three: a head, thorax and abdomen. The head contains a pair of compound eyes and a proboscis. The proboscis is kept curled, and unfurled when the butterfly is drinking juices from plants, animal matter or even mud. A butterfly consumes food only in liquid form. The antennae have scent organs and are used for smelling food as well as locating mates. Some butterfly species possess a special patch of scales called androconia, where pheromones are manufactured. These are wafted into the breeze to help attract mates.

The thorax contains three pairs of legs. In some butterflies the first pair is not fully formed and is not used for locomotion. The legs have a hip joint, or coxa, and a femur, tibia and tarsus. The tarsi (feet) have taste buds. A butterfly therefore tastes by literally walking on its food.

The abdomen has ten segments, with the last two fused to form the reproductive organs.

LIFE CYCLE OF A BUTTERFLY

Soon after mating the eggs are fertilized. They are stuck onto a leaf with a glutinous liquid that hardens to form a waterproofed membrane. Butterfly eggs appear in a variety of shapes: cylinders, cones, spheres and so on. Depending on the species, a few eggs or over a hundred may be laid. Each butterfly species has a host plant on which it lays eggs. In the conservation of butterflies, understanding the host plants is important.

On hatching, a caterpillar, or larva, is usually tiny and initially simply scrapes the leaf with its mouth. It grows rapidly and soon can be over a hundred times its original body mass. Some caterpillars are camouflaged to look like bird droppings. Caterpillars have ocelli on the head that enable them to detect light and dark, but they cannot see. They can, however, sense vibrations through the host plant. Most caterpillars are leaf eaters. However, a few species (especially in the family Lycaenidae – the blues) are carnivorous, and eat aphids, or ant eggs or ant grubs (see p. 12).

A caterpillar eventually stops growing, attaches itself by a thread of silk to a plant and sheds its larval skin. Over the next few hours it changes into a pupa inside a chrysalis. Inside, one of the miracles of life takes place as the pupa metamorphoses into a butterfly. The metamorphosis can be delayed if suitable conditions are not present for the adult butterfly.

The adult emerges by splitting the chrysalis and hangs upright until fluid is pumped into the wings, which stiffen and dry out. After a few tentative flaps to test its wings, the adult flies off to feed and mate.

BUTTERFLY TOPOGRAPHY

THE PARTS OF A BUTTERFLY

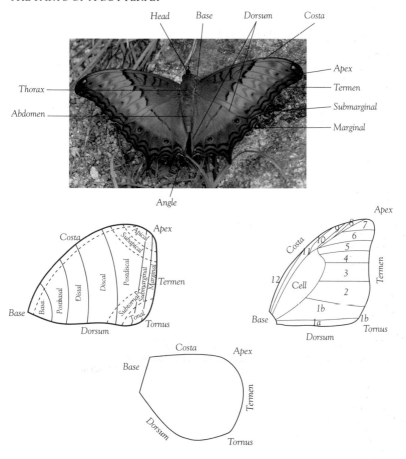

Feeding

The adult butterfly feeds on nectar and is able to utilize a wider choice of plants than its caterpillars (see below) can feed on. An endemic butterfly will almost certainly depend on one or more native plant species as its food plant. It is unlikely that its caterpillars will be able to adapt to a plant that is introduced. On the other hand, provided its proboscis can reach the nectar, an adult butterfly readily adapts to introduced species. The invasive *Lantana* is an example of a plant that has benefited many skippers.

Trap lining Butterflies use a method of feeding called trap lining. They visit plants in flower in a sequence, and at times revisit the same plant and flower. On the former logging road in Sinharaja, which provides the main access from Kudawa, I often see certain butterfly species trap lining. These include some of the swallowtails such as the Red Helen and Blue Mormon.

Protein As most butterflies drink nectar they do not have a source of protein. The absence of protein is a key reason why most adult butterflies live from only a few days to a few weeks. The longwings (family Heliconidae) in South America ingest pollen and take in protein. They are thus able to live for up to eight months.

Mud sipping Butterflies also need nutrients that are not available from nectar. Mud sipping is the term given to butterfly activity at salt licks: usually damp patches of ground from which they ingest minerals such as phosphorus. In the dry zone, in places like Yala, clouds of whites and yellows (family Pieridae) can gather at ephemeral water puddles. In Morapitiya and Sinharaja, mud-sipping sites are often the best places to see butterflies like the Five-bar Swordtail and Painted Sawtooth. Perspiration attracts butterflies for the same reason – my attempts to photograph lineblues in Morapitiya have often been frustrated by the butterflies perching on my sweaty hands. Many butterflies also take minerals from animal scat and rotting fruits. High canopy-dwelling species such as the Common Nawab are best seen when they settle on fruit or salt licks.

Food plants for caterpillars Butterflies, or more specifically in their larval stage the caterpillars, are host specific. This means that a butterfly's caterpillars will feed on the leaves of just one or a few plant species. If the plant is not found in a particular area, that butterfly will not be found there. For example, the beautiful Banded Peacock lays its eggs on the Satinwood Tree (*Burutha* in Sinhalese). This is common in the dry zone, and the distribution of the tree in the dry zone matches the distribution of the butterfly. A common butterfly in cities such as Colombo is the gorgeous Common Jezebel. It lays its eggs on the parasitic *Loranthus*. This plant has spread because it parasitizes the mango tree, which is widely cultivated in home gardens. The Common Jezebel has followed suit.

Butterfly Behaviour

Butterflies employ a variety of behaviours, some of which are specific to certain species or groups.

Ant associations in blues Butterflies in the family Lycaenidae, known as the blues, have a fascinating relationship with ants. The larvae of the blues are often taken underground into ants' nests by specific species of ant that look after them and tend to them. This means providing them with protection from predators as well as food. Thus in this case the caterpillar is not a herbivore, but a carnivore eating ant larvae. The ants tolerate this. However, there is no free lunch for the larvae, and the reward for the ants is a sweet secretion exuded by the larvae on which the ants feed. As an aside, ants also have associations with plants that have extrafloral necataries which provide sugary secretions. The ants provide protection for the plants by stinging potential predators of the plants (that is, herbivores).

False eyes Many insects, including butterflies in both the adult and larval stages, have eyes that they suddenly reveal to predators to startle them into thinking that they have mistakenly attacked a much larger animal. Many of the blues have eyes permanently on the hindwings which are constantly in view. Some, like the silverlines, have a thin projection on the hindwing. When viewed from behind, the tail end looks like a head complete with eyes and antennae. The theory here is that predators such as birds will instinctively attack what they perceive to be the area of the head with 'eyes'. This allows a butterfly to escape with damaged rear wings. If you look closely at butterflies you can often see that some show what looks like a tear from a bird bite slicing off a part of the rear wing. If you look closely, you will notice that many of the small blues rub their wings together all the time. This could be an adaptation to make the eyes on the wings look like a moving head. Butterflies that are cryptically coloured when their wings are folded may, when pecked at, suddenly reveal bright colours on the upperwings. This could serve to startle predators in the same way that false eyes do.

Cannibalism Butterfly caterpillars are known to eat each other in more than one family. The blues seem more predisposed to cannibalism than other butterflies. The Apefly is a

Common Jays mud sipping

tiny species of blue common in Sri Lanka, which can be seen even in Colombo's gardens. However, it is often overlooked because it is small and more white than blue. The larvae of the blues are carnivorous, feeding on aphids and ant grubs. This may make the larvae more likely to cannibalize on larvae of their own species. An advantage of cannibalism is that the survivors have less competition when feeding on their host plants.

Hair pencils Male butterflies that are ready to mate attract females by advertising their presence using pheromones dispersed like a scent in the wind. Most butterflies use scent brands on the wings. Species such as the Common Indian Crow evert, from the tip of the abdomen, 'hair pencils' that waft pheromones. If you see a Common Indian Crow fluttering about with a tuft of yellow on its abdomen this is what is happening.

Hill topping After emergence, butterflies often fly up hills. This has been seen all over the world and is known as hill topping. I have noticed it in both the Knuckles and Horton Plains areas in Sri Lanka, where thousands of whites and yellows that have emerged stream past heading uphill. This may be a strategy that has evolved for the two sexes to meet at the top to mate. This spectacular movement of butterflies could have given rise to Adam's Peak being given the alternative name of Samanala Kanda, or Butterfly Mountain.

In the 1980s, when I was still at school, I remember a day when a huge ribbon of butterflies flew overhead, covering a strip of sky. Even at this time pesticides were being used intensively in the countryside. I have never again seen this phenomenon – where a patch of sky was covered with butterflies – in Sri Lanka. I can imagine how incredible a spectacle may have been witnessed by observers decades or centuries earlier. However, even now I believe large movements of butterflies take place, although they may not be as spectacular because the numbers are less and are diluted over vast distances. In November 2001 I noted in my journal that as I drove past Uda Walawe National Park, from Timbolketiya to Tanamanwila, thousands of whites and yellows were streaming across the road in a northerly direction. Hill topping may arise as a result of local movements or migrations, but is not to be confused with migrations across physical boundaries like oceans and political boundaries.

Migration During migration, butterflies may cross physical barriers like oceans to move across political boundaries. When whale watching off Kalpitiya, Trincomalee and Mirissa, I have often encountered butterflies from different families. One of the most conspicuous is the Common Rose – one of the swallowtails – as well as whites (family Pierdiae). In the case of dragonflies it has been shown that certain species are genetically predisposed to fly in a certain direction after emergence. This forces the gene pool to mix and also helps a species to survive by colonizing new habitats. A similar genetic mechanism may operate in butterflies – hence a species flying south from India will reach Sri Lanka, but successive generations of its progeny flying south from Mirissa may perish in the sea. I have also wondered whether some of the Painted Ladies I have seen in Horton Plains and even in the Kotte Marshes near Colombo are migrants. The Painted Lady is a species famous in Europe for large migrations, which take place from the coastal regions of North Africa into Europe. On 24–25 May 2009, an estimated 28 million butterflies crossed into the south of the UK.

Territoriality It may surprise people to know that butterflies exhibit territorial behaviour. Some of the best-known studies of this were done on a European species, the Speckled Wood, which frequents woodland rides. During summer I often see Speckled Woods patrolling the footpaths at the London Wetland Centre in London. Studies have shown that a territory holder has a home advantage. If one is captured and released into another's territory, its chances of winning are less than when it is confronting intruders on its own territory. The Nymphalidae (the brush-footed butterflies) are famous for their territorial behaviour, and they will lunge at people, dogs and even passing cars from their perches. In May 2008 I saw a Great Eggfly guarding a territory for three days by chasing away other animals that flew past – a rather dangerous ploy as other fliers included birds that eat it. I have also been confronted by nymphalids, including a Blue Admiral at Tangamalai Sanctuary near Haputale in December 2012. It flew at me and buzzed me a few times. Later it calmed down and either took a liking to me or thought my shoulder made a useful perch. It was a magical moment to walk alone on a path through montane forest with a Blue Admiral on my shoulder.

Mimicry There are two forms of mimcry, Batesian and Müllerian. In Baetsian mimicry a non-poisonous species imitates a poisonous species. In Müllerian mimicry two poisonous species mimic each other's patterns to reinforce the message that an animal with that pattern is poisonous. Many Müllerian mimics are in the same family.

A good example of Müllerian mimicry is found in the Danaidae family (the tigers). Butterflies in this family ingest poisonous glycosides from their food plants and store them in the adult body, making themselves distasteful to predators. They advertise this by being marked strongly. Thus all of the tigers, such as the Common Tiger, Dark Blue Tiger and Ceylon Tiger (a montane endemic), advertise a common association by wearing the same pattern on their wings. Mimicry is a topic that has fascinated evolutionary biologists, and for several decades a debate raged since Henry Bates first propounded it from his observations in Amazonia.

The Müllerian mimics are genuinely distasteful, and it is in their interest to advertise this and reinforce a common message by being boldly patterned in tiger stripes. Protected by their toxins, they fly about in a lazy, slow flight. The Tree Nymph is a good example of a slow-flying tiger. Both in flight and at rest, the wing striping shows well, warning predators that they taste bad. Of course a few die as young predators have to kill and eat a few before they associate tiger-striped butterflies with a bad taste.

The Batesian mimics, such as the Common Mime (a swallowtail), Common Palmfly (a satyrid) and Dark Wanderer (a pierid), do not taste bad but do mimic the tigers. Note that the three Batesian mimics mentioned here are from three different families, mimicking the members of one family, the tigers. For this ploy of mimicry to work, the 'models' they are mimicking, in this case the tigers, must be more plentiful. The more Batesian mimics there are, the more they will dilute the signal to predators that tiger stripes signal a bad taste. In fact, if the Batesian mimics became more abundant, the meaning of the signal would reverse. Various studies in Africa found that the Batesian mimics are about 20 per cent of the models. So a four out of five chance of a predator eating a bad-tasting tiger seems to be a good quotient to maintain the value of the tiger stripes as a bad taste signal.

The Batesian mimics often fly faster than the model. By the time a butterfly flies past and a bird realizes that it was a Batesian mimic, the potential victim has made its escape. Batesian mimics that taste good also seem to prefer to use camouflage on the underwing when they are at rest. For example, the finely barred brown underparts of the Common Palmfly make it disappear when it lands. As the female flies it looks like a Common Tiger, which it mimics. Some butterflies that are models for mimicry are polymorphic: that is the male or female may have forms that look different. Batesian mimics also evolve multiple forms to mimic the different forms of the models.

Sri Lanka's Butterfly Fauna

A total of 247 species of butterfly and skipper has been recorded from Sri Lanka. The Great Crow *Euploea phaenareta*, Blue Glassy Tiger *Ideopsis similis*, Orange Migrant *Catopsilia scylla*, Banana Skipper *Erionota torus* and Yellow Palm Dart *Cephrenes trichopepla* are believed to be accidental introductions. One of the most intriguing butterflies is Green's Silverline, described to science in 1896. There were no records of it since the first specimen was collected, and some subsequent authors of butterfly books removed it from the Sri Lankan list because they thought that the original description of the species had been based on a variation of other silverline species. On 13 March 2008 naturalist Nadeera Weerasinghe photographed a female blue laying eggs near World's End and sent me the pictures. Unable to identify it from the literature I had, I passed it on to George van der Poorten, who realized it was the long-lost endemic Green's Silverline.

George and Nancy van der Poorten in *The Butterfly Fauna of Sri Lanka* published in 2016 recognised the following species as endemic to Sri Lanka. Further revisions are expected.

Danaidae (Milkweed Butterflies)
Ceylon Tiger *Parantica taprobana*
Ceylon Tree Nymph *Idea iasonia*
Satyridae (Satyrs or Browns)
Ceylon Palmfly *Elymnias singhala*
Ceylon Treebrown *Lethe daretis*
Ceylon Forester *Lethe dynaste*
Tamil Bushbrown *Mycalesis subdita*
Cingalese Bushbrown *Mycalesis rama*
Jewel Four-ring *Ypthima avanta*
Nymphalidae (Brush-footed Butterflies)
Blue Oakleaf *Kallima philarchus*
Lycaenidae (Blues)
Woodhouse's Four Lineblue *Nacaduba ollyetti*
Pale Ceylon Six Lineblue *Nacaduba sinhala*
Ceylon Cerulean *Jamides coruscans*
Milky Cerulean *Jamides lacteata*

Ceylon Hedge Blue *Udara lanka*
Singalese Hedge Blue *Udara singalensis*
Clouded Silverline *Spindasis nubilis*
Green's Silverline *Spindasis greeni*
Ormiston's Oakblue *Arhopala ormistoni*
Ceylon Royal *Tajuria arida*
Pieridae (Whites and Yellows)
Lesser Albatross *Appias galena*
One-spot Grass Yellow *Eurema ormistoni*
Papilionidae (Swallowtails)
Ceylon Rose *Pachliopta jophon*
Common Birdwing *Troides darsius*
Hesperidae (True Skippers)
Black Flat *Celaenorrhinus spilothyrus*
Tricolor Pied Flat *Coladenia tissa*
Hedge Hopper *Baracus vittatus*
Ceylon Ace *Halpe ceylonica*

Rare Ace *Halpe egena*

Tropic Dart *Potanthus satra*

Decorated Ace *Thoressa decorata*

Paintbrush Swift *Baoris penicillata*

Harish Gaonkar who was working on a taxonomic revision of the butterflies of the Oriental region has noted that the following 19 species are endemic to the Western Ghats and Sri Lanka.

Long-brand Bushbrown *Mycalesis visala subdita*
Glad-eye Bushbrown *Nissanga patnia junonia*
White Four-ring *Ypthima ceylonica*
Baronet *Symphaedra nais*
Tawny Rajah *Charaxes psaphon*
Striped Pierrot *Tarucus nara*
Ceylon Silverline *Spindasis ictis ceylanica*
Scarce Shot Silverline *Spindasis elima*
Redspot *Zesius chrysomallus*
Monkey Puzzle *Rathinda amor*
Common Jezebel *Delias eucharis*
Striped Albatross *Appias libythea libythea*
White Orange-tip *Ixias marianne*
Little Orange-tip *Colotis etrida limbatus*
Blue Mormon *Papilio polymnestor parinda*
Common Banded Peacock *Papilio crino*
Crimson Rose *Pachliopta hector*
Spotted Small Flat *Sarangesa purendra*
Chestnut Angle *Odontoptilum ransonnettii*

BUTTERFLY CONSERVATION IN SRI LANKA

Status Unlike in countries like the UK, for example, there is no butterfly-recording scheme in place in Sri Lanka. However, George van der Poorten has made a best-estimate assessment of the status. The results published in the National Red List 2012 of Sri Lanka are summarized below. As can be seen, 125 of a total 247 species (just over half) merit some degree of concern.

Status	No. of species
Critically Endangered (CR)	21
Endangered (EN)	38
Vulnerable (VU)	40
Near Threatened (NT)	20
Data Deficient (DD)	6
Total	**125**

In Sri Lanka butterflies continue to be in decline due to loss of forest cover. Furthermore, the extensive use of insecticides kills tens of millions of butterflies each year at stages from egg to caterpillar to adult. The large butterfly swarms reported decades earlier are now gone. Everybody can play a part in helping to conserve butterflies by planting some of their larval food plants in a corner of their home or office garden.

Conservation Three things can help to bring back butterflies.
1. Protecting existing forest cover and helping to regenerate degraded forests.
2. Reducing the use of insecticides.
3. Planting larval food plants for the caterpillars, and encouraging the growth of patches of wild flowers to support the adults.

Larval Food Plants

As mentioned in the section on butterfly behaviour, each butterfly species has one or a few plants (typically in the same plant family) that are required for the caterpillar to feed on. The part eaten is usually the leaves, but at times it can be the flowers or parts of the flowers such as the anthers. The larval (caterpillar) food plant that a butterfly needs is given where known in the majority of the species accounts that follow. From a conservation angle, knowledge of the food plants helps people to grow the right plant to bring butterflies back to their home and office gardens, whether in the cities or suburbs.

At the time of going to press, documented evidence of the food plants in Sri Lanka has only been published for some species. The most reliable references are the recent and on-going work of George and Nancy van der Poorten. An earlier reference that is also useful is the three volumes of the *Lepidoptera of Ceylon* published by Frederick Moore in 1880–1881. For many species the authors of books have based their assumptions about larval food plants on field work conducted in India. Although a plant species found in India may be rare or absent in Sri Lanka, plants used in India do at least indicate which plant family or genus is a likely candidate in Sri Lanka.

For some of the swallowtails, tigers, whites and yellows, nymphalids and blues, the confirmed use of a plant has been published in the papers by the van der Poortens. They in turn attribute some of the confirmed use in Sri Lanka to personal communications with others. This book is aimed at beginners and to avoid making it unwieldy each source is not cited in the text. For this, the more specialist reader will need to read the original publications by, for example, Moore and the van der Poortens. For many of the species, information on the larval food plants is derived from a variety of sources, and in turn derived by them from older literature. The main other sources of information for the food plants have been Woodhouse, Ormiston, d'Abrera and Gamage. Woodhouse in turn drew from previous authors, and later authors have drawn from Woodhouse.

Plant Families used by Larvae

Listed below are the plant families used by the larval stage of butterflies mentioned in the text against the corresponding butterfly family. For the purpose of this listing, the older and traditional family classification has been retained, where the Danaidae, Satyridae, Amathusidae, Acraeidae and Libytheidae are treated as separate families rather than subfamilies of the Nymphalidae. The respective subfamily treatment would be to regard them as the subfamilies Danainae, Satyrinae, Morphinae, Acraeinae and Libytheinae in the family Nymphalidae. Note that one plant family may be used by several families of butterfly.

As only some of the species in a family are covered in this book, it is not a full list of the plant families used by that butterfly family in Sri Lanka. Bear in mind that in other countries, a butterfly family may have species that use plants from families not on this list. Therefore this table is only indicative. However, it does give an idea of the range of plant families used by different butterfly families. To attract butterflies to a garden, you need nectaring plants that are favoured by the adult stage of butterflies. However, also – critically – the larval food plants are needed. This is especially the case if a butterfly garden is to be created as a mini nature reserve for their conservation.

Fifty plant families are listed below, and this illustrates the botanical challenge confronting any lepidopterist in the tropics. To have a good understanding of the life cycle of butterflies, you need to know the plants, and in the tropics, where so many species of plant and butterfly are found, this presents an enormous challenge. Butterfly watching is greatly enhanced by a knowledge of plants

Danaidae	Apocynaceae, Asclepiadaceae, Boraginaceae, Moraceae and Periplocaceae
Lycaenidae	Acanthaceae, Anacardiaceae, Boraginaceae, Combretaceae, Crassulaceae, Cycadaceae, Dioscoreaceae, Dipterocarpaceae, Euphorbiaceae, Fabaceae, Hippocrateaceae, Loranthaceae, Malpighiaceae, Mimosaceae, Myrtaceae, Plumbaginaceae, Rhamnaceae, Rubiaceae, Rutaceae, Sapindaceae, Smilacaceae and Symplocaceae.
Papilionidae	Annonaceae, Aristolochiaceae, Fabaceae, Lauraceae, Magnoliaceae, Rutaceae and Verbenaceae.
Satyridae	Arecaceae and Poaceae.
Amathusidae	Arecaceae.
Nymphalidae	Acanthaceae, Anacardiaceae, Dipterocarpaceae, Ebenaceae, Euphorbiaceae, Fabaceae, Flacourtiaceae, Gesneriaceae , Loranthaceae, Malvaceae, Melastomataceae, Meliaceae, Mimosaceae, Passifloreceae, Rhamnaceae, Rubiaceae, Smilacaceae, Tiliaceae, Urticaceae, Verbenaceae and Violaceae.
Acraeidae	Cucurbitaceae and Passifloreceae.
Riodinidae	Myrsinaceae.
Pieridae	Brassicaceae, Capparaceae, Loranthaceae and Salvadoraceae.
Libytheidae	Urticaceae.
Hesperiidae	Arecaceae, Dioscoreaceae, Fabaceae, Poaceae, Smilacaceae, Sterculiaceae and Zingiberaceae.

THE SPECIES ACCOUNTS

A standard structure has been used in the species accounts. The many gaps show how little is presently known about Sri Lanka's butterflies. In some cases the categories for sexes, flight and food plants have been deliberately omitted because my personal observations or what is published in the available literature is insufficient for that information to be included. For each species, wingspan measurements are given after the butterfly's common and scientific names at the beginning of every account. Endemics are identified with the symbol ℮ after the scientific name.

Description Unlike in the case of birds, there is very little literature on the identification of difficult species of butterfly not just in Sri Lanka, but generally outside the developed countries in, for example, Europe and North America. I have attempted to point out differences between similar species. In many cases I have had to invent terms such as 'sausages' in silverlines, 'peaks and troughs' in the Ceylon Treebrown, and so on. The identification of features is based mainly on my photographs, and published illustrations in Woodhouse and plates in d'Abrera. However, given the variability from one individual to another, the features cited may not always be reliable. Over time, as more people study butterflies, for some species reliable field characteristics may become better known.

Many butterfly species have wet-season and dry-season forms. In Sri Lanka both the wet-season and dry-season forms may be on the wing at the same time. Unless they are likely to cause confusion, I have not mentioned the colour variations in the seasonal forms.

Sexes Some butterfly species show different external appearances for adult males and adult females. Some of these sexual differences are shown. As Sri Lankan butterflies are studied better, more information may emerge on separating the sexes. Some of the older literature (like *The Fauna of British India including Ceylon and Burma: Butterflies*, by C. T. Bingham) describes small variations in the structure of the wing, for example, between the sexes. As these can only at best be used in the hand, I have not included many of these details.

Flight Descriptions such as those pertaining to flight can be subjective and vary from observer to observer. To be consistent on the descriptions of flight, I have used the information in John and Judy Banks' *A Selection of the Butterflies of Sri Lanka* together with my own observations. Where I have not made any specific notes, this section is absent. However, for many species the family has a characteristic flight. For example, all the blues have the habit of a rapid zigzag flight when they are searching for a food plant.

Distribution This is based heavily on accounts by Woodhouse and d'Abrera supplemented by my own observations. For skippers, as very little is documented on the distribution, for several species I have noted a few sites from where I have confirmed records.

Food plants For more details, see p. 18. Generally, the larval stages of butterflies feed on one or more host plants that is specific to that species. An exception is the larvae of the blues, most of which are carnivorous (see p. 12).

The food plants of the adult and larval stages are given where known. In some species plants are only known or conjectured to a generic level. In others the food plants are not known at all. For example, in the case of the Dark Blue Tiger the larval food plants in Sri Lanka had not been published until 2011, when George and Nancy van der Poorten produced a paper. The Dark Blue Tiger is not a rare species and this demonstrates how much even basic information on the Sri Lankan lepidoptera is still in the process of being recorded.

For many species the adult food plant is often not given due to my lack of familiarity with some of the plants, although I have seen many of these species sipping nectar, at times from introduced plants. The vernacular English '[E:]' or Sinhala '[S:]' name, where known, of the food plant is shown in brackets. The name of the family is generally given in parentheses, for example '(family Asclepiadaceae)'.

Status The general status of butterflies is not to be confused with the National Red List Status, which uses IUCN criteria. The status used here is based on an unpublished checklist shared with me many years ago by Priyantha Wijesinghe. The treatment of a species as an endemic is based on details shared by Harish Gaonkar from his book under preparation, *The Atlas of the Butterflies of the Western Ghats and Sri Lanka*.

Larvae The larvae are not covered in this book. Readers could refer to d'Abrera, Woodhouse and others, and the technical papers by George and Nancy van der Poorten.

Nomenclature A Sinhala and Tamil nomenclature for butterflies has not yet been developed. It is hoped that this book will serve to highlight the urgent need for a nomenclature that is suitable for common usage, but is also scientifically robust, like the Sinhala nomenclature developed for birds by D. G. A. Perera and Sarath Kotagama.

GLOSSARY

Androconial patch Specially modified scales on upperwing of a butterfly that help to disperse pheromones. They help to attract females and may also be used to warn other males.
Brand The androconial scales (see above) often form a distinctive patch or line on the wing, referred to as a brand.
Cryptic Hidden. In this book term usually refers to cryptic patterns or colouration for camouflage. A cryptic species is one that is hidden to the casual viewer because it is similar to another species and may not show up as another species unless analyses using molecular techniques are carried out.
Mud puddling/mud sipping Habit of certain species of butterfly of congregating on the ground in areas that are damp with water or animal urine. They absorb mineral nutrients from the soil through their probosces.
Seasonal dimorphism Many butterfly species show a different external appearance between those that hatch in the dry season and those that do so in the wet season. Generally the wet-season form is marked more heavily or is darker.
Strigae Refers to bands on underwings of certain species of blue.

DANAIDAE (MILKWEED BUTTERFLIES OR TIGERS)
This family gets its name from the larval host plants, which are in the milkweed family. They are also known as tigers and crows. The name tiger derives from the striped patterns found in many of the species. Their forelegs are modified into brushes and cannot be used for walking. This supports their inclusion as a subfamily in the Nymphalidae, or brushfoots. Many species extrude 'pencils' from the anus. These are believed to be used to waft pheromones to attract females (see p. 14). The caterpillars feed on plants that result in toxic alkaloids being ingested, which makes the adult butterflies distasteful. This results in the adults being 'models' for species that mimic them (see p. 15).

Glassy Tiger ▪ *Parantica aglea aglea* 70mm

DESCRIPTION Looks similar to other blue tigers. Cell pattern on under hindwing and 'prongs of a fork' pattern towards dorsum edge of under hindwing help to distinguish it from other species. Underwing can look very brown, depending on how the light catches the wings. **SEXES** Similar. **FLIGHT** Low and slow flying. Wingbeats rapid and deep, giving a fluttering impression. **DISTRIBUTION** Found up to highlands. Most common in southern half of the island. **LARVAL FOOD PLANTS** *Tylophora tenuissima*, *T. indica*, *Heterostemma tanjorense* (family Asclepiadaceae).

Ceylon Tiger ■ *Parantica taprobana* 85mm ⓔ

DESCRIPTION Black and white on upperwings; brown underwings. Unlikely to be confused with any of the other tiger species. **SEXES** Similar. **DISTRIBUTION** Found above 1,000m. May be seen in gardens around Nuwara Eliya, as well as along roadside verges in forested areas, for example at Horton Plains, Bomurella and Hakgala. Also found in submontane forest clearings in the Knuckles massif. **LARVAL FOOD PLANTS** *Cynanchum alatum*, and possibly *Tylophora cordifolia* and *T. indica* (family Asclepiadaceae).

Male

Female

Blue Glassy Tiger ■ *Ideopsis similis exprompta* 60–75mm

DESCRIPTION Similar in appearance to Blue Tiger (see p. 24). Two lines in cell of under hindwing break cell into three segments. This helps to separate it from similar Blue Tiger. **SEXES** Similar. **FLIGHT** Slow and lazy; occasionally glides. To cover distance, flies with a few flaps of wings followed by long glide with wings held horizontal. **DISTRIBUTION** Confined to south-west of the island. **LARVAL FOOD PLANTS** *Tylophora indica* (family Asclepiadaceae). Has been seen laying eggs on *Parsonia alboflavescens* (family Apocynaceae), but it has not been confirmed that larvae use this as a food plant.

Blue Tiger ■ *Tirumala limniace exoticus* 90mm

DESCRIPTION In upper forewing, near dorsum edge, there is a large amount of misshapen blue. **SEXES** In female cell on under hindwing is not broken by any lines: it is an almost triangular wedge of light blue. Male has a dark line in under hindwing cell. **FLIGHT** Slow. **DISTRIBUTION** Found throughout the island, but most common in lowlands. **LARVAL FOOD PLANTS** *Wattakka volubilis* [S: Anguna] (family Asclepiadaceae). The Van der Poortens observed that larvae refuse to feed on *Calotropis gigantea* [S: Wara]. **ADULT FOOD PLANTS** *Lantana* sp.

Dark Blue Tiger
■ *Tirumala septentrionis musikanos* 70–85mm

DESCRIPTION Three sets of 'V's on under hindwing are distinctive. Blue noticeably darker than blue of Blue Tiger (see above), which it may be seen swarming with on plants. **SEXES** Similar. **FLIGHT** Slow. **DISTRIBUTION** Found throughout the island, but most common below 1,000m. Often swarms with other Danaidae species on *Heliotropium strigosum* and *H. indicum*. **LARVAL FOOD PLANTS** *Heterostemma tanjorense* (family Asclepiadaceae). In laboratory conditions larvae have fed on *Wattakka volubilis* [S: Anguna] (family Asclepiadaceae), but its use in the wild has not been confirmed. **ADULT FOOD PLANTS** *Heliotropium strigosum* and *H. indicum*

Common Tiger ▪ *Danaus genutia genutia* 75–85mm

DESCRIPTION Can be distinguished from Plain Tiger (see below) by its boldly marked dark veins. Rear edge of underneath of hindwing has two 'white chains'.
SEXES Similar. **FLIGHT**
Slow, but rapid when evading predators. **DISTRIBUTION**
Found throughout the island to 1,200m. Most common in coastal areas. Common butterfly of roadside verges and gardens.
LARVAL FOOD PLANTS
Oxystelma esculentum [S: Usepale], *Cynanchum tunicatum* [S: Kan-kumbala] and *Tylophora tenuissima* (family Asclepiadaceae).
Different food plants may be used depending on distribution of food plants on the island. **ADULT FOOD PLANTS** *Heliotropium indicum* (family Boraginaceae).

Plain Tiger ▪ *Danaus chrysippus chrysippus* 60–80mm

DESCRIPTION Lacks bold black vein markings of Common Tiger (see above). On hindwings it has one, not two, 'chains'. **SEXES** Female occurs in two forms. One
has dark and white subapical markings on under forewing.
The other has a plain forewing with a dark border. **FLIGHT** Slow.
DISTRIBUTION Found throughout the island.
LARVAL FOOD PLANTS
Asclepias currasavica [Kan-Kumbala], *Calotropis gigantea* [S: Wara], *Gomphocarpus physocarpus* and *Pentatropis capensis* (family Asclepiadaceae).
ADULT FOOD PLANTS
Heliotropium indicum (family Boraginaceae).

Great Crow ■ *Euploea phaenareta corus* 95–100mm

DESCRIPTION Large size readily distinguishes this species from Common Indian Crow (see opposite) and **Double-branded Crow** E. *sylvester*. Male can be distinguished from female by bowed dorsum of forewing. On upper forewing apex a double arc of white spots

is present. Upperwing rich chocolate-brown, darker than that of other 'brown crows'. FLIGHT Slow. DISTRIBUTION Confined to south-west of the island from Negombo to Galle, which coincides with the distribution of its food plant. There is also a population beside the paddy fields on the logging road used by visitors in the Sinharaja reserve from the Kudawa entrance. It is thought that it may have been introduced when Gon-kaduru was planted near the paddy fields. Manders (1904) has suggested that this butterfly may have been accidentally introduced from China into the port of Galle, which may explain its current distribution. LARVAL FOOD PLANTS *Cerbera odollam* [S: Gon-kaduru] (family Apocynaceae).

Brown King Crow
■ *Euploea klugii sinhala* 80–90mm

DESCRIPTION Male's androconial patch on under forewing can show as a patch of violet. Of the 'brown crows' this is the species that can show violet on upper forewing in fresh specimens. Female lacks white spot on hindwing cell that is found in female of Common Indian Crow (see opposite), **Double-branded Crow** E. *sylvester* and Great Crow (see above). SEXES In male dorsum of forewing is deeply bowed. FLIGHT Slow and lazy. Habit of perching high, well above head height. DISTRIBUTION Throughout lowlands, ascending to 700m. Most common in dry lowlands. LARVAL FOOD PLANTS Toothbrush Tree *Streblus asper* [S: Geta netul] (family Moraceae).

Common Indian Crow ■ *Euploea core asela* 85–95mm

DESCRIPTION Double-branded Crow *E. sylvester* is very similar. Male can be told apart by Double-branded Crow having a single, velvety black brand near dorsum edge of upper forewing. Whitish streak on underside. Female has only whitish streak. In Double-branded, male has two brands on dorsum edge of upper forewing, and both male and female have two whitish streaks on under hindwing. In Common Indian Crow, and in Great Crow (see opposite), under hindwing does not have white spot on interspace 7. However, Double-branded Crow and Brown King Crow (see opposite) do have this white spot. Bear in mind that there can be some variation between individuals in the presence or absence of spots. **SEXES** Female lacks androconial brand on upperwing. **FLIGHT** A few wingbeats followed by fast glide with wings held in 'V' shape. Can

move surprisingly quickly in this 'flap-and-glide' mode when it needs to. **DISTRIBUTION** Found throughout the island. Common butterfly of roadside verges and gardens. **LARVAL FOOD PLANTS** *Ficus religiosa* [S. Bo], *F. pumila*, *F. benjamina* (family Moraceae), *Gomphocarpus physocarpus*, *Pentatropis capensis* (family Asclepiadaceae), *Cryptolepis buchananii*, *Hemidesmus indicus* (family Periplocaceae), *Nerium oleander* [S. Kaneru], *Adenium obesum*, *Allamanda cathartica* [S: Wal-ruk-attana], *Parsonsia alboflavescens*, *Ichnocarpus frutescens* and *Ochrosia oppositifolia* (family Apocynaceae).

Ceylon Tree Nymph ■ *Idea iasonia* 120–150mm ⓔ

DESCRIPTION Unmistakable butterfly with slow, sailing flight and black-and-white pattern. **SEXES** Similar. **FLIGHT** Slow, with intermittent glides where it seems to hang suspended in the air.
DISTRIBUTION Endemic. Found in lowlands and mid-hills where good stretches of forest remain. Most common in wet zone. Appears to prefer areas close to streams. The stretch of paddy fields at Sinharaja beside the access from the barrier (from the Kudawa entrance) is a reliable location. **LARVAL FOOD PLANTS** *Parsonsia alboflavescens* (family Apocynaceae).

SATYRIDAE (SATYRS OR BROWNS)
The satyrs are generally a nondescript colour and are shade-loving species. Their forelegs are brush-footed and not suitable for walking, and their flight is weak. Some species occur in dry- and wet-season forms. Their food plants are in the families Graminae and Palmaceae.

Common Evening Brown ▪ *Melanitis leda leda* 60–65mm

Wet-season form

Dry-season form

DESCRIPTION Very similar to Dark Evening Brown (see opposite). Best distinguished by forewing apex having a better defined hook in Dark Evening Brown. There is a lot of variability between the wet-season and dry-season forms. In the dry-season form there are 'eyes' on under hindwing. Wet-season form has dark bands against a brown background. **SEXES** Similar. **FLIGHT** Rapid and jerky. **DISTRIBUTION** Found throughout the island. Often enters buildings, where it is seen hanging onto walls. During the day it can be encountered perfectly camouflaged on leaf litter. **LARVAL FOOD PLANTS** *Oryza sativa* [S: Wi, Uru wi], *Imperata cylindrica*, *Setaria barbata*, *Panicum maximum*, *Leersia hexandra*, *Cenchrus ciliaris* and *Ischaemum timorense* [S: Rila rat tana] (family Poaceae).

Dark Evening Brown
■ *Melanitis phedima tambra* 60–70mm

DESCRIPTION Distinguished from similar Common Evening Brown (see opposite) by a pronounced hook forming on termen near apex. **SEXES** Male noticeably darker than female on upperwings. **FLIGHT** Jerky and close to the ground. **DISTRIBUTION** Found in lowlands to mid-hills. **LARVAL FOOD PLANTS** *Setaria barbata, Ischaemum timorense* [S: Rila rat tana], *Digitaria didactyla* [E: Blue Grass] and *Axonopus compressus* [S: Potu-tana, E: Carpet Grass] (family Poaceae).

Common Palmfly ■ *Elymnias hypermnestra fraterna* 60–80mm

DESCRIPTION Medium-sized brown butterfly. Male brown above with yellowy-brown trailing edge to upper hindwing. Underside marbled with pale striations. Usually not far from a palm tree. Despite being permanently present in my home garden in Colombo, after six years I had not seen it perching even once with its wings open. **SEXES** Female in flight very similar to Plain Tiger (see p. 25) in orange-yellow; a complete contrast to male, which is drab on upperparts. Female also more heavily patterned on underside and shows some traces of white towards apical area of forewings. **FLIGHT** Brisk; does not like to fly. Settles very quickly. **DISTRIBUTION** Common to 457m. A home garden with even a single palm tree may attract this butterfly. **LARVAL FOOD PLANTS** *Livistona sp., Arecha catechu, Calamus thwaitseii, Cocos nucifera* [S: Pol, E: Coconut], *Phoneix pusilla, Caryota Urens* [S: Kitul, E: Fishtail Palm, Kitul Palm], *Cyrtostachys renda* and *Chrysalidocarpus lutescens* (family Arecaceae).

Male

Female

Ceylon Treebrown ■ *Lethe daretis* 45–55mm ⓔ

DESCRIPTION This species, and Tamil Treebrown (see below) and Ceylon Forester *L. dynaste*, can be identified by closely observing basal part of under forewings. This contrasts with paler outer half, separated by pale band between the two halves in the three treebrowns and Ceylon Forester. In Tamil Treebrown edge of basal half (in discal area) has pronounced 'peaks and troughs'. In Ceylon Treebrown under

hindwing circles all look more or less 'solid', unlike in Tamil Treebrown and Ceylon Forester. **SEXES** Female has a pale, diagonal band on upper forewing. **FLIGHT** Rapid and somewhat wavy, at around head height. I have found it hard to keep track of individuals that have engaged in brief flights along a path before settling down at head height. **DISTRIBUTION** Found in highlands at 1,700–2,700m. Occurs near clumps of bamboo. Cloud forests in Nuwara Eliya district are good locations for this butterfly, and Horton Plains National Park is certain to yield sightings of it. **LARVAL FOOD PLANTS** *Sinarundinaria debilis* (family Poaceae) is the only known food plant, and it is absent from forests where this plant is absent.

Tamil Treebrown ■ *Lethe drypetis drypetis* 55mm

DESCRIPTION Very similar to **Ceylon Forester** *L. dynaste*. Both species have eyespots on under hindwing, with two white-centred solid spots and the others 'fragmented' with

sprinkles of 'star dust'. In Ceylon Forester the two solid circles are distinctly bigger than the other circles. **SEXES** Male's upper forewings uniformly brown. Female has pronounced pale band in distal area, and smaller pale band, or path, in apical area. **DISTRIBUTION** Found throughout the island subject to high rainfall and presence of bamboo. The conditions make it most likely to be encountered in hill country. **LARVAL FOOD PLANTS** *Bambusa multiplex* [S: Cheena Una, E: Bamboo] and *Dendrocalamus giganteus* (family Poaceae).

Common Bushbrown ■ *Mycalesis perseus typhlus* 38–55mm

DESCRIPTION Similar to **Tamil Bushbrown** M. *visala subdita* and **Dark-brand Bushbrown** M. *mineus*, but can be distinguished from those species by the four lowest ocelli (eyespots) on under hindwing being arranged in curved rather than straight line. **SEXES** Similar. **FLIGHT** Can fly very fast and low when it needs to – so fast that it is hard to keep track of. **DISTRIBUTION** Common throughout the island. Often seen in somewhat shaded, shrubby vegetation. **LARVAL FOOD PLANTS** *Oryza sativa* [S: Wee, E: Rice], *Axonopus compressus* [S: ptu-tana, E: Carpet Rice] and *Leersia hexandra* (family Poaceae).

Dark-brand Bushbrown ■ *Mycalesis mineus polydecta* 45mm

DESCRIPTION Distinct eyespot in both upper forewing and upper hindwing. Upperwing eyespot patterns comprise a pale spot with concentric dark smudged circle and pale outermost circle. **SEXES** Female paler on upperwing than male. **DISTRIBUTION** Found in lowlands to mid-hills. **LARVAL FOOD PLANTS** *Panicum maximum* [E: Guinea Grass] and *Axonopus compressus* [S: Potu-tana, E: Carpet Grass] (family Poaceae).

Cingalese Bushbrown ■ *Mycalesis rama* 45mm ⓔ

DESCRIPTION Easiest to identify by its underwing. Hindwing has seven distinct eyespots, of which three have thick black circles. Both under forewing and under

hindwing have a broad pale stripe in middle. Under forewing has two large eyespots. **SEXES** Female may be paler on upper surface than male, but this is to be confirmed. **FLIGHT** Can be fast in moving from one perch to another. Seems reluctant to fly long distances. **DISTRIBUTION** Found in lowland rainforests. Appears confined to lowland forests where bamboo is found. May not be as rare as early literature suggests. **LARVAL FOOD PLANTS** Probably one or more species of bamboo. *Ochlandra stridula* [S: Bata] is a likely species.

Glad-eye Bushbrown ■ *Mycalesis patnia patnia* 35–40mm

DESCRIPTION Distinctive underwing pattern. Seldom seen with upperwings open. Wet-season form richer in colour than dry-season form. **SEXES** Similar. **FLIGHT** Slow; keeps close to the ground and flight is not sustained for long. **DISTRIBUTION** Found throughout the island but less common in north. Prefers to perch low in dense shrubs. **LARVAL FOOD PLANTS** *Isachne globosa* [S: Bata-della] and *Cyrtococcum trigonum* (family Poaceae).

Medus Brown

■ *Orsotriaena medus mandata* 45–55mm

DESCRIPTION Very distinctive butterfly with strong white band on underwings and eyespots. **SEXES** Similar. Female slightly larger than male, with more rounded forewings. **FLIGHT** Slow; keeps close to the ground and flight is not sustained for long. **DISTRIBUTION** Found mainly in southern half of the island, to 1,500m. Most active during the evening. **LARVAL FOOD PLANTS** *Oryza sativa* [S: Wee, E: Rice], *Panicum maximum*, *Leersia hexandra* and *Axonopus compressus* (family Poaceae).

White Four-ring ■ *Ypthima ceylonica* 30–35mm

DESCRIPTION Common butterfly that can be seen in the city as well as the countryside. Under hindwing largely pale with only a sprinkle of brown mottling. Under forewing has 'four rings', and large eye with blackish centre fringed with pale ring and another small eyespot. Eyespot pattern is repeated on upperwings, with upper forewing brown and upper hindwing sharply demarcated in white in discal area. Sufficiently distinct to avoid confusion with any other species. **SEXES** Similar. **FLIGHT** Rapid flutter, usually close to the ground. In flight can look smaller than it is, because brown on wings is not as visible as white. As a result in flight it looks like a small white butterfly, about the size of the smaller Psyche (see p. 78). **DISTRIBUTION** Found throughout the island to 1,200m. Common and found in grassy areas in home gardens. **LARVAL FOOD PLANTS** *Axonopus compressus* and *Cyrtococcum trigonum* (family Poaceae). Larvae feed at night.

AMATHUSIDAE
There is only one species of this family in Sri Lanka. The forelegs of Amathusidae butterflies are underdeveloped, their flight is not strong and they are shade loving. Their food plants are in the families Graminae and Palmaceae.

Southern Duffer ▪ *Discophora lepida ceylonica* 70mm

DESCRIPTION Large butterfly. Male is dark, with upper forewing having two pale irregular spots and upper hindwing having blackish spot. Female has bluish-grey and broad broken band on upper forewing. Upper hindwing has three broken bands, with middle band made up of concave markings. **SEXES** Kehimkar (2008) notes that females fly during the day and males are active at dusk and dawn. **FLIGHT** Jerky flight; settles down quickly. **DISTRIBUTION** Limited to bamboo forests in wet lowlands. Confined to Western Ghats and Sri Lanka. **LARVAL FOOD PLANTS** Unknown, but the species has been seen laying eggs on *Ochlandra stridula* [S: Bata] (family Poaceae).

NYMPHALIDAE (BRUSH-FOOTED BUTTERFLIES)

In these butterflies the first pair of legs cannot be used for walking, hence their common name. They are medium- to large-sized butterflies that are often seen basking in the sun. Many species have eyespots on the wings, and a few have tails. In several there are distinct differences in the male and female (sexual dimorphism). The group has a wide range of behavioural attributes, from feeding (for example, some are necatar feeders, while others are attracted to ripe fruits and/or animal droppings), to being migrants or non-migrants. Some members in the group, such as the Painted Lady, are known for spectacular migrations in Europe.

Angled Castor ▪ *Ariadne ariadne minorata* 45–50mm

DESCRIPTION Very similar to Common Castor (see p. 36). Common Castor has crenulated wings in both female and male, and in this species dark crenulated lines in basal and discal areas are 'paired'. In Angled Castor the lines are spaced apart. SEXES In Angled Castor, only female shows crenulation. Male can be distinguished from female by having top third unmarked on under hindwing. FLIGHT Slow, erratic flight. DISTRIBUTION Common throughout the island to about 915m. Found on verges, degraded cultivation and similar places. Common name refers to larval host plant – the castor-oil plant. LARVAL FOOD PLANTS *Ricinus* sp. and *Tragia* spp. (family Euphorbiaceae).

Common Castor ▪ *Ariadne merione taprobana* 45–60mm

DESCRIPTION Zigzag line parallel to termen on upper forewing. The next band on distal area is a chain of 'popcorn' shapes with dark centres. Angled Castor (see p. 35) lacks

'popcorn' shapes. **SEXES** In the field they look similar. **FLIGHT** Weak and jerky. Flies low over the ground. Settles on ground-level herbage. **DISTRIBUTION** Found island-wide, but least common at high elevations. **LARVAL FOOD PLANTS** *Ricinus* sp. and *Tragia* sp. (family Euphorbiaceae). **ADULT FOOD PLANTS** *Ricinus* sp. and *Tragia* sp. (family Euphorbiaceae).

Joker ▪ *Byblia ilithyia* 50mm

DESCRIPTION Hindwing is dark edged on termen, with rectangular shapes formed by black veins joining a thick, broad black band running parallel to termen. Forewing has black band along costa, and two more broad bands penetrating wing arising from dorsum. Each band is irregular. Wet-season form is richer in colour than dry-season form. **SEXES** In female black markings are dark brown. **DISTRIBUTION** Found in drier, more arid areas of Sri Lanka. Most likely to be encountered in north. **LARVAL FOOD PLANTS** *Ricinus* sp. and *Tragia* sp. (family Euphorbiaceae).

Indian Fritillary ■ *Argynnis hyperbius taprobana* 60–70mm

DESCRIPTION Upperwing superficially similar to that of Common Leopard (see below), but Indian Fritillary has a blue band on upper hindwing termen. Underwing reminiscent of that of **Painted Lady** *Vanessa cardui*, but this has at least four well-marked ocelii ('eyes'). Overall impression is of orange butterfly on upperparts with underparts richly patterned. Both sexes have distinct upper hindwing pattern of thick black edge with pair of white bands of opposite-facing, concave crescents. In both sexes upperwings are flecked with black spots, and one rectangular black hollow marking on upperwing near costa. **SEXES** In female outer half of upper forewing is dark with conspicuous white band. **FLIGHT** Strong, wavy flight. **DISTRIBUTION** Found in highlands at elevations above 1,500m. Can be seen sunning itself on lawns of Hakgala Botanical Gardens and on roadside verges of Horton Plains (for example). Considered mainly a Palearctic species, surviving as a relict species in temperate habitats in the tropics. **LARVAL FOOD PLANTS** *Viola patrinii*, a cultivated plant (family Violaceae).

Common Leopard ■ *Phalanta phalantha phalantha* 60mm

DESCRIPTION May be confused with male Indian Fritillary (see above) as forewing pattern is similar. However, Indian Fritillary has blue stripe on rear edge of upper hindwing, and shows at least one 'rectangle' on upper forewing. **SEXES** Similar. **FLIGHT** Rapid. Often flies high, above head height. **DISTRIBUTION** Found throughout the island to 2,000m. Occurs in a mixture of habitats, from open areas to gardens and forests with clearings. **LARVAL FOOD PLANTS** *Flacourtia sp.* (family Flacourtiaceae) and *Smilax* sp. (family Smilacaceae).

Tamil Lacewing ■ *Cethosia nietneri nietneri* 90mm

DESCRIPTION Large butterfly with strongly wavy-edged wings, the so called 'lacewing'. Overall impression of upperparts is of a strongly marked black-and-white butterfly. Underwing has broad orange band parallel to termen edges of wing. Upperwings and underwings marked with spots, bars and waves. **SEXES** Female duller than male and lacks orange diffusion on basal areas of upperwings. **FLIGHT** Flies with deep and rapid wingbeats; almost moth-like. Fairly straight flight. **DISTRIBUTION** Found from lowlands to around 1,300m. **LARVAL FOOD PLANTS** Passifloracea family species.

Cruiser ■ *Vindula erota asela* 100mm

DESCRIPTION Large size, 'swallowtail' points on tornus of hindwing, and two large eyespots on hindwing surfaces. **SEXES** Male orange with thin black band cutting across middle of wing surfaces, and two thin, wavy bands parallel to outer edges of wings. Female has the same markings, but upper surface is a mixture of brown outer edges, shading into blue with pale irregular band in middle of upper forewing. **FLIGHT** Strong, purposeful flight. I have seen it at times very high up in trees. **DISTRIBUTION** Occurs throughout the island, but is scarce. **LARVAL FOOD PLANTS** *Adenia palmata* and *Modecca* sp. (family Passifloraceae).

LEFT: *Male*; RIGHT: *Female*

Tamil Yeoman ■ *Cirrochroa thais lanka* 60mm

DESCRIPTION Upperwings and underwings similar to those of **Rustic** *Cupha erymanthis placida*. In Rustic, black on upperwing curves more broadly, and comes in about a third of the way from apex. In Tamil Yeoman black is not as extensive. Forewings of Rustic are also rounded. In Tamil Yeoman ternum is falcate (bowed in). **SEXES** Female has a more pronounced falcate apex on forewing than male, and has a transverse discal stripe, which is weak or absent in male. **FLIGHT** Fast and straight. **DISTRIBUTION** Found throughout the island to 1,300m. Often occurs in and around forest patches. **LARVAL FOOD PLANTS** *Hydnocarpus wightiana* (family Flacourtiaceae).

Blue Admiral ■ *Kaniska canace haronica* 60–70mm

DESCRIPTION Medium-sized butterfly with contrasting, very broad light blue band in middle across upperwings. Wing edges are wavy and tornus is a little pointed. Underwing is cryptic and offers perfect camouflage. Fearless butterfly that will fly out and buzz even people who walk past its territory. I have had it buzz me and then perch on my shoulder – a magical moment. **SEXES** Similar. **FLIGHT** Fast and fairly straight with only a hint of zigzagging. Can be confused with **Common Bluebottle** *Graphium sarpedon teredon* if a good look is not had as it whizzes past. **DISTRIBUTION** Found in mid-hills up to highlands. Prefers habitats with dense forest cover adjoining fast-flowing streams. **LARVAL FOOD PLANTS** *Smilax prolifera* (family Smilacaceae).

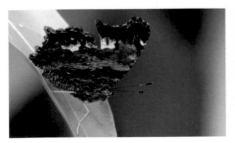

Indian Red Admiral ■ *Vanessa indica nubicola* 60mm

DESCRIPTION Boldly patterned on upperwings, lending the impression of a black and reddish-orange butterfly. Distal half of upper forewing black with broken band of white spots. Discal area has reddish-orange band, and basal area is brown. Upper hindwing is brown with orange border on termen. **SEXES** Female duller than male. **FLIGHT** Purposeful; a little bouncy. **DISTRIBUTION** Found in highlands above 1,300m. Seen basking on lawns in gardens in hills. **LARVAL FOOD PLANTS** *Girardinia heterophylla* (family Urticaceae).

Chocolate Soldier

■ *Junonia iphita iphita* 70–85mm

DESCRIPTION Overall impression is of a chocolate-brown, medium-sized butterfly. There are three diffuse darker brown bars on upper forewing meeting costal edge at right angles. Indistinct, small spots parallel to termen. Upper hindwing has distinct spots parallel to termen. Irregular open, oval squiggles on costal edge. **SEXES** Similar. **FLIGHT** Stiff winged. Does not fly far, and alights on the ground and on flowers. Confiding, allowing a close approach.

DISTRIBUTION Found throughout the island. Sometimes seen mud puddling. Often basks with wings open. **LARVAL FOOD PLANTS** *Strobilanthes* sp. (family Acanthaceae).

Blue Pansy ■ *Junonia orithya patenas* 40–50mm

DESCRIPTION Upperwings are diagnostic – a deep blue (except at apices) with red circles. Underwing is pale with some marbling. Under hindwing is vaguely similar to that of other pansies such as Grey, Lemon and even Peacock Pansies (see pp. 42–3). However, under forewing has two distinctive orange-yellow bands bordered with black, and a broad, diffuse white band bordered with black on one side. None of the nymphalids found in Sri Lanka have this pattern. A butterfly perched with its wings closed is hard to see. In flight it can look like a butterfly with dark and light upperparts and underparts respectively. On landing it has a habit of flickering its wings open a few times, before closing them in a vertical position. **SEXES** Female has dark area on basal part of upper hindwing. **FLIGHT** Swift and short. It seems to almost leap away and land somewhere else. I have watched this butterfly fly out from a Balu Nakuta (*Stachytarpheta* sp.) flower stalk, whirl around and fly back to another stalk of flowers. It does this so fast that it is hard to follow the movements with your eyes. In rapid flight blue of upperwing does not show and butterfly looks pale brown. **DISTRIBUTION** Historical distribution was throughout the island, but it seems to be very scarce now. I have only seen it around the Vil Uyana hotel in Sigiriya, but am aware of records from the Kandyan hills. **LARVAL FOOD PLANTS** Family Acanthaceae. **ADULT FOOD PLANTS** *Stachytarpheta indica* [S: Nil-nakuta] (family Verbenaceae).

Lemon Pansy ■ *Junonia lemonias vaisya* 45–60mm

DESCRIPTION Not likely to be confused with any other species. Red-rimmed eye on each wing and brown forewings with clear white flecks are quite distinctive. **SEXES** Similar. **FLIGHT** Swift. **DISTRIBUTION** Common throughout the island to 700m. Often seen in abandoned chenas and other degraded habitats. **LARVAL FOOD PLANTS** Family Acanthaceae.

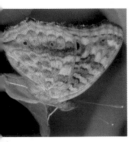  

Peacock Pansy ■ *Junonia almana almana* 40–65mm

DESCRIPTION Orangish wings, large 'eyes' on upper hindwing, and bars on costal (leading) edge of forewing are distinctive. Underwing very pale and superficially similar to that of Grey Pansy (see opposite), but Peacock Pansy has more prominent 'eyes' on underwing. **SEXES** Similar. **DISTRIBUTION** Common throughout the island to 1,000m, although distribution extends to highest mountain plateau. **LARVAL FOOD PLANTS** D'Abrera quotes *Gloxinia* spp. (family Gesneriaceae). The genus does not occur naturally in Asia, but the flowers are found in horticulture, and the Gesneriaceae family has many genera in Sri Lanka. Also *Osbeckia* spp. (family Melastomataceae) and species of family Acanthaceae.

Grey Pansy ■ *Junonia atlites atlites* 55–60mm

DESCRIPTION Pale grey butterfly with a row of ocellii on forewings and hindwings.
Underwings are even paler, and rear hindwing has a thin diagonal line going across it.
SEXES Similar. **FLIGHT** Fast. **DISTRIBUTION** Found throughout the island to 1,000m.
Common butterfly in disturbed habitats. **LARVAL FOOD PLANTS** Family Acanthaceae.
ADULT FOOD PLANTS Often visits flowers of invasive *Lantana* (family Verbenaceae).

Blue Oakleaf ■ *Kallima philarchus* 80–90mm ℮

DESCRIPTION Large butterfly that is surprisingly hard to see because its underwing
camouflages it so well. It visits rotting fruits and is worth looking out for in feeders
at Sinharaja and in some of the lodges around Kithulgala. Forewings and hindwings
curve gracefully into a point. Upperwings shade from mixture of brown and blue in
hindwings, to bright blue on basal and discal half of forewings darkening on forewing
distal area. **SEXES** Female browner than male on hindwings, and pale band in middle
of upper forewing is white. **DISTRIBUTION** Occurs to 700m. Found locally in forests
where *Strobilanthes* species, which are its food plant, are found. I have found it to be
scarce. **LARVAL FOOD PLANTS** *Strobilanthes* sp. (family Acanthaceae). **ADULT FOOD
PLANTS** Fond of feeding on rotting fruits.

Male

Great Eggfly ■ *Hypolimnas bolina bolina* 65–110mm

DESCRIPTION Upperparts similar to those of smaller male Danaid Eggfly (see below), but in Great Eggfly pale wing-patches are more bluish and small. In Danaid pale patches on upperwing may not show bluish shine. Larger of the two pale patches is a large oval and much more prominent than large pale patch in Great Eggfly. Danaid has distinctive underwings with white bands against orange. Under hindwing has broad pale band, and white of upper forewing can also be seen in under forewing. There is no confusion between the two species when underwing pattern is seen. **SEXES** Female mimics both forms of Plain Tiger (see p. 25). **FLIGHT** Swift and bouncy. Habit of circling around as if it is undecided. Males will aggressively sally forth at other animals that fly past a territorial perch. **DISTRIBUTION** Found throughout the island. **LARVAL FOOD PLANTS** *Strobilanthes* spp. (family Acanthaceae), *Elatostema cuneatum* and *Fleurya interrupta* (family Urticaceae).

Danaid Eggfly ■ *Hypolimnas misippus* 70–85mm

DESCRIPTION Male has similar wing pattern to Great Eggfly's (see above), but the pale wing-patches are relatively larger in Danaid Eggfly. Trailing edge does not show

two parallel lines of arrowheads like in Great – just one line of arrowheads. Female Danaid copies Plain Tiger (see p. 25), showing both forms of its model. Plain Tiger has two forms, one with black subapical markings and another without them. In Plain Tiger rear edge of hindwing has white spots neatly enclosed in black, allowing the 'model' to be distinguished from a mimic. **SEXES** See above. **FLIGHT** Fast. **DISTRIBUTION** Found throughout the island to 1,000m. **LARVAL FOOD PLANTS** Species in families Acanthaceae and Malvaceae.

Common Sailor ■ *Neptis hylas varmona* 45–60mm

DESCRIPTION Upperparts similar to Chestnut-streaked Sailor's (see below). Species can always be told apart by Common Sailor having two broad white bands on hindwing and Chestnut-streaked Sailor having only one. Under hindwing also shows this difference. In Common Sailor upperwing lacks hint of brown. **SEXES** Both sexes show variation in richness of colour of underwing. **FLIGHT** Slow and stiff winged. **DISTRIBUTION** Common throughout the island. **LARVAL FOOD PLANTS** Species in families Fabaceae, Malvaceae and Tiliaceae.

Chestnut-streaked Sailor ■ *Neptis jumbah nalanda* 60mm

DESCRIPTION This species can be told apart from similar Common Sailor (see above) by the one broad white band on the upper hindwing – Common has two broad white bands. Under hindwing has row of dark spots in median area. **SEXES** Females have thin third white band on upper hindwing. **FLIGHT** Has two modes of flight. In one it engages in short, sallying flights where wingbeats are rapid and almost quivering but very shallow, lending a stiff-winged flight. When travelling from one patch of foliage to another it uses a mixture of rapid, shallow wingbeats and a stiff-winged glide. During the glide wings are held horizontal. Sometimes banks a little while engaging in glides. **DISTRIBUTION** Found throughout the island to 1,500m. Occurs in a variety of habitats from forests to village gardens. **LARVAL FOOD PLANTS** Species in families Fabaceae, Malvaceae and Rhamnaceae.

Male

Female

Common Lascar ■ *Pantoporia hordonia sinuata* 40–50mm

DESCRIPTION Orange and brownish-black colouration superficially similar to Joker's (see p. 36), but in Lascar bands are straight, unlike in Joker. Latter is more like Common Sailor (see p. 45), with white being replaced by orange. Underwings lack distinct stripey pattern of Common Sailor, and are cryptic with diffused bands. **SEXES** Similar. **FLIGHT** Gentle and sailing. **DISTRIBUTION** Found to 1,300m.
LARVAL FOOD PLANTS *Acacia* spp. and *Albizia* spp. (family Mimosaceae).

Commander ■ *Moduza procris calidasa* 60mm

DESCRIPTION Distinctive butterfly. Upperwings have broad white 'V'-shaped band in middle of wings against a darker background. On close inspection patterning on

upperwing is more complex, with red-and-black bands on basal area as well as in area between edge of wing and white band in middle. Termen is crenualte. Upperwing pattern mirrored on much paler underwing. **SEXES** Similar, but female is bigger than male. **FLIGHT** Swift. Hunched posture in flight with wings held at an angle below horizontal (forming an inverted 'V' shape). Clipper (see opposite) has identical posture, but is much larger. Covers ground rapidly with powerful down strokes of wings. **DISTRIBUTION** Found to 1,800m but most common in mid-hills. Most likely to be seen where good-quality forest patches remain. **LARVAL FOOD PLANTS** *Mussaenda frondosa*, *Wendlandia* sp. and cultivated *Cinchona* sp. (family Rubiaceae).

Clipper ■ *Parthenos sylvia cyaneus* 90mm

DESCRIPTION Broad white 'splashes' in forewing against generally blue-green upperwing make confusion unlikely if a good view is obtained. Will occasionally mud sip. Bodhinagala is a good site for both Clipper and Commander (see opposite). **SEXES** Similar. **FLIGHT** Purposeful. Can be very fast at times. Hunched posture in flight with wings held at an angle below horizontal (forming inverted 'V' shape). A Clipper patrolling a forest path reminds me of a 'straight-flying' insectivorous bat. It can rapidly ascend vertically 10–15m into the canopy. **DISTRIBUTION** Found to 1,000m. Most likely to be seen where good-quality forest patches remain. **LARVAL FOOD PLANTS** *Modecca* sp. and *Adenia hondala* [S: Hondala, Potahonda] (family Passifloraceae).

Redspot Duke ■ *Dophla evelina evelina* 80mm

DESCRIPTION Habit of alighting on foliage that is above head height. As a result much of butterfly is obscured by leaf it perches on. This factor, combined with its green colour, often results in it being often overlooked, or hard to photograph even when seen. **SEXES** Similar. **FLIGHT** Very swift. Hard to track in flight. **DISTRIBUTION** Found to 500m. D'Abrera says it is found near rivers and waterways, but I have not noticed a particularly strong connection with flowing water habitats. **LARVAL FOOD PLANTS** *Anacardium occidentale* (family Anarcardiacea), *Disopyros candolleana* and *D. melanoxylon* [S: Kadumberiya] (family Ebenaceae).

Baronet ▪ *Symphaedra nais* 50–60mm

DESCRIPTION Overall impression is of a medium-sized orange butterfly with brown markings. May be confused with Joker (see p. 36), but upper hindwing has neat spots on trailing edge and termen is neatly bordered by brown. Upper forewing has brownish-black bars, but they lack the irregular shapes of those in Joker. **SEXES** Similar. **FLIGHT** Flies low, close to the ground. I have seen it perching on the ground with wings open. **DISTRIBUTION** D'Abrera states that this species has a very local distribution along the south-east coast up to around Trincomalee in the north-east. In Uva province it is seen in good numbers at Nilgala. **LARVAL FOOD PLANTS** In India reported to use *Shorea robusta* (family Dipterocarpaceae). **ADULT FOOD PLANTS** *Diospyros* spp. (family Ebenaceae).

Gaudy Baron ▪ *Euthalia lubentina psittacus* 48–60mm

DESCRIPTION Female could, if looked at superficially, be overlooked for more common Commander (see p. 46) because of white band on upper forewing. Male lacks broad white band and only has two rows of small white spots on forewing. Female has big white shapes that at a distance look like a broad white band. Both sexes have two rows of red spots on hindwings. When wings are closed lack of white band on hindwing makes for a pronounced difference from Commander. A restless butterfly, even when it is mud sipping it does not stay in one place for long. It comes to bait and decaying fruits. **SEXES** Male does not have broad broken white band on forewing. **FLIGHT** Fast; fairly straight. **DISTRIBUTION** Found throughout the island to 1,500m. Not common anywhere. Rare north of central highlands. **LARVAL FOOD PLANTS** Plants in family Loranthaceae. **ADULT FOOD PLANTS** Sips nectar from Durantha (an introduced garden plant).

Female

Baron ■ *Euthalia aconthea vasanta* 55–80mm

DESCRIPTION A concave termen, a line of spots near edge of hindwing, misshapen rectangles on cell of forewing and hindwing, and absence of any bright-coloured patches help identify this species. However, males can show a green gloss on upperwings. **SEXES** Sexes are different, with male having glossy green upperwings and female being brown with white band on upper forewing. **FLIGHT** Fast, jerky, curving flight often above head height. A Baron once hit my car's windscreen on the Chilaw-Puttalam Road. We pulled over and placed it gingerly on foliage and photographed it. After 5–10 minutes it suddenly flew away, showing how robustly it is built. **DISTRIBUTION** Found throughout the island to about 1,000m. Can be overlooked in home gardens as it may keep to upper reaches of fruiting trees such as Jak. **LARVAL FOOD PLANTS** *Mangifera indica* [S: Amba, E: Mango] and cultivated *Anarcardium occidentalis* [S: Cadju, E: Cashewnut] (family Anacardiaceae).

Male   *Female*

Common Nawab
■ *Polyura athamas athamas* 50mm

DESCRIPTION Underwing has a roughly oval, green shape that is distinctive. Species frequents tree tops and is seen only rarely when it descends to mud sip. **SEXES** Similar in pattern and colour; larger female has broader wings. **FLIGHT** Fast. **DISTRIBUTION** Absent in far north; widespread to 1,500m. I have only seen it where there is still good forest cover. **LARVAL FOOD PLANTS** *Aglaia roxburghiana* [S and E: Mahogany] (family Meliaceae).

Tawny Rajah ■ *Charaxes psaphon psaphon* 90mm

DESCRIPTION Large butterfly unlikely to be confused with any other. Rufous upperparts with forewing edged broadly in black. Underwing is quite cryptic. **SEXES** Differ, with female having broad white band on forewing. Female also more contrastingly patterned on underwings than male. **FLIGHT** Fast and bounding. **DISTRIBUTION** Found in lowlands and to mid-hills. Rarely seen because of its habit of staying inside dense forest. **LARVAL FOOD PLANTS** *Aglaia roxburghiana* [E: Mahogany] (family Meliaceae). **ADULT FOOD PLANTS** Feeds on fermenting fruits and dead animal matter.

Male   *Male*

Black Rajah ■ *Charaxes solon cerynthus* 75mm

DESCRIPTION Double 'swallowtail', and underside has prominent white band and a few jagged black lines. Unlikely to be confused with any other species. Habit of perching a few feet above the ground and sallying forth to challenge other males or intercept females. Can be found visiting faeces of animals, from which

it absorbs minerals. **SEXES** Similar in pattern, but female is slightly larger. According to Woodhouse, females have longer tails than males. **FLIGHT** Swift. **DISTRIBUTION** Found throughout the island, below 700m. Probably very scarce, as such a large butterfly would be noticed if it was common. My observations and those of others suggest that the North Central Province is one of the best areas in the country in which to find it. **LARVAL FOOD PLANTS** *Bauhinia racemosa* [S: Maila] and *Tamarindus indica* [S: Siyambala] (family Fabaceae).

ACRAEIDAE
In Sri Lanka there is only one species of this family, which is centred in the Afrotropical region. The Acraeidae are small to medium-sized butterflies, in colours ranging from yellow to red, with long forewings. The family is mainly centred around Africa; there are around 200 species, with an extension to Asia. Another centre of distribution is in South America. The Acraeidae butterflies synthesize cyanide compounds and are distasteful to predators. Where there is an abundance of these species, they are mimicked by other species.

Tawny Coster
▪ *Acraea terpsicore* 50–65mm

DESCRIPTION Elegant butterfly; the only member of its family found in Sri Lanka. White 'pearl chain' set against black on rear edge of hindwing is distinctive. Forewings relatively long compared with hindwings. **SEXES** Similar. **FLIGHT** Slow and measured, but can be fast when disturbed. **DISTRIBUTION** Found throughout the island. **LARVAL FOOD PLANTS** *Modecca* spp. (family Passifloraceae) and family Cucurbitaceae.

> **LIBYTHEIDAE**
> This family gets its name from the long palpi, or 'beaks', of the butterflies. Unusually in this family, the males have forelegs that are not properly formed for walking, but in the females they are normal. All members of the family have the same wing shape. Only two species are found in Sri Lanka

Club Beak ▪ *Libythea myrrha rama* 45–55mm

DESCRIPTION Very similar to **Beak** *L. celtis lepitoides*. Male Club Beak darker on upperparts than Beak, and orange and dark brown contrast more strongly. Less space between orange stripe on middle of forewing and circular patch at its end compared with Beak, where 'rod and ball' is less of a snug fit. Two adjoining pairs of pale spots on forewing. The one furthest from base and lower of the two is a tighter fit in Club Beak than in Beak. **SEXES** Orange bands on forewing broader in female than male. **FLIGHT** Very swift. Settles on the ground, where it is easily overlooked. **DISTRIBUTION** Found throughout the island; slightly more prevalent in wet-zone forests than in other areas. **LARVAL FOOD PLANTS** *Celtis* spp. (family Uritcaceae).

RIODINIDAE
This family is at its most numerous in the neotropics, but is also found in the Oriental and Australasian regions. The butterflies have a robust build and hold up their wings in a manner reminiscent of the blues (family Lycaenidae). Only one species is found in Sri Lanka. The Riodinade are believed to have split from the Lycaenidae. They are associated with ants. Very little is known of the ecology of the species found in Sri Lanka.

Plum Judy ■ *Abisara echerius prunosa* 40–50mm

DESCRIPTION Distinctive habit of swivelling in jerky movements after it has perched. Wings are held slightly open, similar to many skippers. There are variations in size, and in wet- and dry-season forms. However, confusion with another species is unlikely. In Rummassala I have noticed that this is one of the few butterflies that is active during the late evening. **SEXES** Male darker and more boldly marked than female. Fresh males have purple iridescence on upperparts. **FLIGHT** Alternates between slow satyrid flight to fast, skipper-like dashes in the evening. **DISTRIBUTION** Found throughout the island to 150m. **LARVAL FOOD PLANTS** *Ardisia wightiana* [S: Dan] and *Embelia tsjeriam-cottam* (family Myrsinaceae).

Apefly ▪ *Spalgis epeus epeus* 20mm

DESCRIPTION The small size and impression of a pale fly may result in this butterfly being overlooked for a day-flying moth or some other insect. In flight brown upperparts and pale

patches on discal area in upper forewing do not show. Underparts delicately striated with pattern of thin, wavy lines. The name Apefly derives from a 'monkey-face' pattern on the pupae. **SEXES** Broadly similar, but female is considered to have rounder wings. White patch on upper forewing considered to be more diffuse (but subject to variability in terms of size). **FLIGHT** Erratic; settles quickly. **DISTRIBUTION** Throughout the island to 1,000m. **LARVAL FOOD PLANTS** Larvae are known to be carnivorous and will feed on larvae of mealybugs.

Carnivorous Butterflies Most butterflies in their larval stage (that is the caterpillars) feed on leaves. The blues are in an interesting family where many species have an association with ants (see also p. 53). In the case of many blues, the host ants keep the butterfly larvae in their nests and feed on sugary excretions from the butterfly caterpillars. The caterpillars eat some of the ant pupae and the adult ants do not seem to mind this. The blues also include a subfamily, the Miletinae. Their larvae feed on the larvae of other insects. In scientific parlance they are called entomophagous, meaning that they feed on insects. One of them is the Apefly, whose genus (*Spalgis*) is widespread in the Oriental, Afro-tropical and Australasian regions. Its larvae are important predators of mealybugs, and the larvae also look like the larvae of the mealybugs on which they prey. Species such as Apeflies are an important natural check on the abundance of animals such as mealybugs, which can become pests.

Indian Sunbeam ▪ *Curetis thetis* 40mm

DESCRIPTION Male unmistakable, with its copper upperparts and white underparts. It joins pierids to mud sip, and is easily mistaken for one of the whites and yellows. **SEXES** Female brown on upperparts with pale patches on both pairs of wings. Pale patches bigger on upper forewing. **FLIGHT** Fast, typical pierid flight. **DISTRIBUTION** Throughout the island to 1,300m. Most common near water courses. **LARVAL FOOD PLANTS** Quoted in the literature but unconfirmed for Sri Lanka are *Derris scandens*, *Pongamia glabra* and *Abrus precatorius* (family Fabaceae).

Male

Male

IDENTIFICATION OF LINEBLUES

This is a complex group that is very difficult to identify in the field. Fortunately, only 13 species are currently recognized in Sri Lanka. The advent of digital photography helps enormously as patterns can be compared on a camera or computer, combined with the ability to zoom in on details. Woodhouse suggested that what was in his time 12 species of *Nacaduba* lineblues can be divided into three groups. Three are now in the genus *Prosotas* and one is in the genus *Ionolyce*. The question marks in the tables below are intentional. As field identification improves, it is possible that external field characteristics may be recognized in these species. If not, they will require examination in the hand. The monograph by J. N. Eliot is a useful source of reference.

Pavana Group Basal area of under forewing is *without* white strigae. Tailed.	Other identification features
Large Four Lineblue *Nacaduba pactolus*	Antennae have white tips.
Pale Four Lineblue *Nacaduba hermus*	?
Woodhouse's Four Lineblue *Nacaduba ollyetti*	Black tornal post is elongated.
Berenice Group Basal area of under forewing is with white strigae. Tailed.	
Transparent Six Lineblue *Nacaduba kurava*	Largest, semi-transparent wings.
Pale Ceylon Six Lineblue *Nacaduba sinhala*	?
Opaque Six Lineblue *Nacadube beröe*	Males have dark upperwings.
Dark Ceylon Six Lineblue *Nacaduba calauria*	?
Rounded Six Lineblue *Nacaduba berenice*	Males have pale blue upperwings.
Nora Group On underwings, spaces between strigae are darkened. The three smallest of the *Nacaduba* species. Only Common Lineblue *Nacaduba nora* is tailed.	
Common Lineblue *Nacaduba nora*	Tailed.
Tailless Lineblue *Nacaduba dubiosa*	No tail.
White-tipped Lineblue *Prosotas noreia* (in Woodhouse described as *Nacaduba noreia*).	?
Other Lineblues	
Pointed Ciliate Blue *Anthene lycaenina*	Two tiny tails. In addition to black tornal spot a second black spot on hindwing near costal margin.
Dingy Lineblue *Petrelaea dana*	Two black spots on hindwing margin. Black spots not partially fringed with red.

D'Abrera also suggests a critical examination of the following:

i. On underwings, ground colour and numbers and shapes of bands.

ii. On upperwings, shade of blue and width of dark borders.

iii. In males, shape and length of forewings. The extension of blue from the margins inwards to the base should also be noted, as well as the tint of colour.

Pale Four Lineblue ■ *Nacaduba hermus sidoma* 30mm

DESCRIPTION This is in the Pavana Group, which lacks white strigae in the basal area of the under forewing. Tornal black spot in space 2. Mud sips from moist spots on roads. **SEXES** Female has broad dark margin in costal edge and dark margin on termen. Male has uniform blue upperwings. Woodhouse distinguishes between the two species by describing the strigae on the **Large Four Lineblue** *N. pactolus* as being 'broad, diffuse and continuous and are on an ochraceous background', whereas in Pale Four Lineblue strigae are 'narrow, clearly defined, straight and conjoined compared with *pactolus*'. D'Abrera says male can be separated from Large Four Lineblue by 'narrower pointed forewings', and female by more greenish-blue on upperwings. I suspect most observers will find it hard to discern differences in wing shape. **FLIGHT** Fast and erratic; close to the ground. **DISTRIBUTION** Found in mid-hills. **LARVAL FOOD PLANTS** *Symplocos cochinchinensis* (family Symplocaceae).

Dark Ceylon Six Lineblue ■ *Nacaduba calauria evansi* 20mm

DESCRIPTION This is in the Berenice Group, with white strigae in the basal area of the under forewing. Tailed; forewing has three bands. I have noticed that the topmost band meets the costal margin, then continues down at a sharp angle (near vertically), leaving scallops along costal margin. As patterns can vary a lot, I am not sure how consistent this feature is. Ground colour of wings is dark. Woodhouse did not consider that Dark Ceylon Lineblue and **Opaque Six Lineblue** *B. beröe* can be distinguished without dissection of the genitalia in all but the freshest specimens. However, females of Dark Ceylon have more extensive blue on upperwings than females of Opaque Six. D'Abrera believed that the two lineblues can be separated by Dark Ceylon having a more curved hindwing as it is more distally curved, compared with a more triangular shape in Opaque Six. He notes that 'in *calauria* the angle between the anal margin and the tail is very steeply cut away, whilst on *beröe* it is at right angles to the tail'. **SEXES** Female has broad dark edges on costa and termen. **FLIGHT** Fast and erratic, and close to the ground. **DISTRIBUTION** Uncertain. I have found it in wet lowlands. **LARVAL FOOD PLANTS** Unknown.

Common Lineblue ■ *Prosotas nora ardates* 20mm

DESCRIPTION This is in the Nora Group, where in the underwings the spaces between strigae are darkened. The Nora Group includes the three smallest *Nacaduba* species. In the Nora Group only the Common Lineblue is always tailed, irrespective of the colour form. **SEXES** Extent of blue in female on upper forewings restricted to small triangle in centre extending from base. Dull upper hindwings. Male has uniformly blue upperwings. **DISTRIBUTION** Throughout the island. **LARVAL FOOD PLANTS** *Croton aromaticus* [S: Welkeppitiya] (family Euphorbiaceae), *Terminalia catappa* [S: Kottamba, Kottan] (family Combretaceae), *Acacia caesia* [S: Heenguru Vel], *A. pennata* [S: Hinguru, Godahinguru], *Pithecellobium dulce* [E: Madras Thorn], *Samanea saman* [S: Mara] (family Mimosaceae), *Derris scandens* [S: Ala vel, Bo-kalavel, Kala-wel] (family Fabaceae) and *Allophylus cobbe* (family Sapindaceae).

Tailless Lineblue ■ *Prosotas dubiosa indica* 20mm

DESCRIPTION This is in the Nora Group. Key feature is lack of tail. Other similar species with a single black tornal spot have a tail. Spaces between white strigae are darkened. In

photographs of this species I have seen, it seems that the centres of the strigae have a pale central line, so that the strigae look as though they are clusters of three parallel white lines. **SEXES** Female has relatively narrow strip of blue extending outwards from base, on upper forewing. Upper hindwing is dull. Male has uniformly blue upperwings. **FLIGHT** Fast, low and erratic. Distribution Dry lowlands to foothills of wet zone. **LARVAL FOOD PLANTS** Larvae feed on flowers of *Samanea saman* [S: Mara] and *Albizia odoratissima* (family Mimosaceae).

Angled Pierrot
▪ *Caleta decidia* 26–32mm

DESCRIPTION Somewhat similar to **Common Pierrot** *Castalius rosimon*, but can be separated by its under forewings having a few bold black markings. Angled Pierrot has geometric, right-angled black shapes on wings like a teaching aid for children. Two thick black lines emerge from costa, at right angles to it. Occasionally mud puddles. **SEXES** Similar. **FLIGHT** Fast, low and circling; circles repeated at times. **DISTRIBUTION** Found to mid-hills. Prefers clearings around jungle streams. **LARVAL FOOD PLANTS** *Zizyphus* spp. (family Ramnaceae).

Banded Blue Pierrot ▪ *Discolampa ethion ethion* 20mm

DESCRIPTION Under forewing has two parallel thick black stripes in basal area, which start off joined together on costa. Third thick black line is in distal area, leaving costa at an acute angle. Angled Pierrot (see above) has two thick black lines leaving costa at right angles. One of the easy ways of identifying Banded Blue is by black line near apex, which diverges away from the two parallel lines in basal area. If they were joined at the costa they would make a broad 'V'. When the light strikes termen edges of forewing at a certain angle, it creates a thin blue edge. I suspect this is caused by iridescence off tiny hairs on forewing edges. This is a very beautiful effect that can only be seen under the right conditions and if you are very close. Mud sips occasionally. **SEXES** Very dissimilar. Female has broad black margins to upperwings and underwings, and is similar to female Angled Pierrot. **FLIGHT** Whirling, circling flight, fast and low. Looks all white in flight, and at times silvery-grey. Sometimes a hint of blue on upperwings can be seen in males. **DISTRIBUTION** Found throughout the island, but most common in southern half. **LARVAL FOOD PLANTS** *Zizyphus* spp. (family Ramnaceae).

Male

Dark Cerulean ■ *Jamides bochus bochus* 25–30mm

DESCRIPTION Combination of brilliant metallic blue upperwing, dark underwing and small size helps identify this species. At rest, has a habit of rubbing wings together. Tends to keep wings closed. Under hindwing has a dark tornal spot, with inner half

bordered in orange. Pattern of white lines on underwings is simple. **SEXES** On upper forewing female has a broad dark margin along costa and termen. **FLIGHT** Fast. Bright metallic blue of upperwing is striking. **DISTRIBUTION** Found throughout the island. Woodhouse notes that it is most common in south. **LARVAL FOOD PLANTS** *Derris scandens* [S: Ala vel], *D. elliptica*, *Pongamia pinnata* [S: Magul-karandha], *Tephrosia vogelii*, *Vigna radiate* [S: Mung-eta], *V. unguiculata* S: Me-karal], *Acacia auriculiformis*, *Cajanus cajan* [S: Rata tora], *Gliricidia sepium* [S: Kona], *Pueraria phaseoloides*, *Pterocarpus indicus* and *Tephrosia candida* [S: Boga-medeloa], all in family Fabaceae.

Common Cerulean ■ *Jamides celeno tissama* 27–35mm

DESCRIPTION One of the paler blues. Subtle differences in underwing patterns distinguish it from similar species. Third white band (counting inwards) in under forewing is distinctly broader, a feature shared with **Milky Cerulean** *J. lacteata* but not with **Metallic Cerulean** *J. alecto meillichius*. In Milky Cerulean under forewing's first two white bands on

termen create rhomboid brown shapes within them. In Common and Metallic Ceruleans they are more rectangular in shape. Female is different from male and has a broad dark patch at apex of upper forewing. Usually not difficult to separate from Metallic Cerulean as underwing of Metallic is darker than in Common. **SEXES** On upper forewing, female has broad dark band on termen and broad dark apical wing-tips. **DISTRIBUTION** Found throughout the island. Common species on abandoned plots and other such places. **LARVAL FOOD PLANTS** *Vigna radiate* [S: Mung-eta], *Pongamia pinnata* [S: Ga-karandha], *Pueraria phaseoloides*, *Abrus precatorius* [S: Olinda], all in family Fabaceae.

Forget-me-not ■ *Catochrysops strabo strabo* 30mm

DESCRIPTION Under hindwing has two neat black spots on costal edge. Both sexes have tornal black spot edged with red. **SEXES** Male has unmarked blue upperparts with black tornal spot on hindwing. Female brownish with blue confined to middle of upper forewing. **FLIGHT** Zigzag circling, often close to the ground. **DISTRIBUTION** Found throughout the island to 1,067m. Common butterfly of wasteland, and other open areas and gardens. **LARVAL FOOD PLANTS** *Schleicheria oleosa* (family Sapindaceae).

Male

Female

Pea Blue ■ *Lampides boeticus* 30mm

DESCRIPTION Under hindwing has no black spots other than two distinctive tornal spots. White distal band abutting tornal spots is broad. Broken wavy distal band, which starts from dorsum, makes a sharp kink where it touches dorsum. **SEXES** Similar. **FLIGHT** Strong. **DISTRIBUTION** Ascends up to highest hills, where it is found in open areas. Settles on moist ground. **LARVAL FOOD PLANTS** Family Fabaceae.

Zebra Blue ■ *Leptotes plinius plinius* 20–30mm

DESCRIPTION Both sexes may be confused with Striped Pierrot (see opposite). Latter is confined to area in dry lowlands north of Mannar and Batticaloa. Upperwing surfaces

of the two species very similar; underwing surfaces help to separate them. Striped Pierrot is striped and spotted with black against a white background. Zebra Blue is marbled in varying hues of brown. Mud puddles. **SEXES** Male has plain blue upperwings. Female has dark 'blocks' in discal and distal areas of upper forewing. **DISTRIBUTION** Found throughout the island to highlands, but more common in dry lowlands. **LARVAL FOOD PLANTS** *Plumbago rosea* and *P. zeylanica* (family Plumbaginaceae).

Common Pierrot ■ *Castalius rosimon* 25–30mm

DESCRIPTION Underwing pattern differs from patterns of **Angled Pierrot** *Caleta decidia* and **Banded Blue Pierrot** *Discolampa ethion* ethion, which have fewer and larger black markings against white. Underside of forewing termen has two neat rows of black spots, which are lacking in Angled Pierrot. Frequently mud puddles. **FLIGHT** Generally weak and low, but can at times engage in fast zigzag flight. Looks all white in flight because wingbeats are rapid. **DISTRIBUTION** Found throughout the island to 700m. Prefers open spaces. **LARVAL FOOD PLANTS** *Zizyphus* spp. (family Ramnaceae).

Striped Pierrot ■ *Tarucus nara* 20–25mm

DESCRIPTION For distinguishing this species from similar Zebra Blue, see account for that species (opposite). **SEXES** On upperwing female has pale area in middle of forewing.

DISTRIBUTION Found in dry northern lowlands, north of arc from Mannar to Batticaloa. Occurs on banks of streams and lakes. The known larval food plant is widely distributed in lowlands, so distribution of the butterfly may be influenced by an adaptation to an arid climate, or distribution of specific species of ant that attend it. **LARVAL FOOD PLANTS** *Ziziphus mauritiana* [S: Mahadebara, Debara, Masan] (family Rhamnaceae).

Dark Grass Blue ■ *Zizeeria karsandra* 15–22mm

DESCRIPTION Often basks in the sun. Under forewing has two dark spots in discal areas that separate it from Lesser Grass Blue (see p. 64), which has one spot. **SEXES** Similar, but female is browner than male. **FLIGHT** Weak and low, settling frequently. **DISTRIBUTION** Found throughout the island to 1,500m. Common in grassland. Quick to colonize wasteland, even in cities, once an area has become overgrown with grasses. **LARVAL FOOD PLANTS** *Zornia diphylla* (family Fabaceae).

Male

Female

Lesser Grass Blue ■ *Zizina otis indica* 20–25mm

DESCRIPTION Similar to Dark Grass Blue (see p. 63). Can be told apart from under hindwing pattern of spots. In Lesser Grass Blue and Dark Grass Blue two dark spots lead

away from thorax along costal vein. In Lesser Grass second spot furthest from thorax has another spot below it. The three make a right-angled triangle. Three spots on distal area make another 'cluster'. In Dark Grass spots on costal vein and in distal area can be drawn into a smooth arc. On under forewing spots in distal area of Lesser Grass can be connected neatly into a curved arc. Dark Grass has at least one spot out of step. Therefore as a rule, drawing a neat arc is easier on Lesser's under forewing. **SEXES** Similar, but female is browner than male. **FLIGHT** Weak and low, settling frequently. **DISTRIBUTION** Found throughout the island to highlands. Often basks in the sun. **LARVAL FOOD PLANTS** *Zornia diphylla* (family Fabaceae).

Tiny Grass Blue ■ *Zizula hylax hylax* 15–20mm

DESCRIPTION Pale underneath with distinct black spots. Middle of under forewing has an ochreous arc marking. All other markings are black. Tiny size, as its name suggests. **SEXES** Male is blue on upperparts, female is brownish. **FLIGHT** Rapid wingbeats, but does not cover a lot of ground. Flies low over the ground. **DISTRIBUTION** Common to 600m on grassland. **LARVAL FOOD PLANTS** Species in families Fabaceae and Acanthaceae.

Indian Cupid ■ *Everes lacturnus* 24–26mm

DESCRIPTION Twin orange spots on tornus of under hindwing are distinctive. Small
Cupid (see p. 69) is similar but smaller, as its name suggests. Plains Cupid (see p. 68) has
underwing pattern reminiscent of Indian Cupid's, but underwing is more boldly patterned.
Upperwing of Plains Cupid has broad black margin on forewing. **SEXES** Male a uniform
bright blue on upperwings. Female brownish on upperwings, with diffuse area of light
blue spreading out from bases of forewings. **FLIGHT** Weak and low. **DISTRIBUTION**
Found throughout the island to highlands. Common in disturbed habitats with grassland.
LARVAL FOOD PLANTS Plants in family Fabaceae.

Male

Red Pierrot ■ *Talicada nyseus nyseus* 30–38mm

DESCRIPTION Distinctive small butterfly with red border to under hindwing.
Upperwings brown with large orange band on rear of hindwing. The species is hardly
ever seen basking with wings open. **SEXES** Similar in pattern, with female being
slightly larger than male. **FLIGHT** Zigzag, whirly flight low over the ground. Settles
frequently. **DISTRIBUTION** Found throughout the island. Visits flowers even in
town gardens. **LARVAL FOOD PLANTS** *Kalanchoe calycinum* (family Crassulaceae).

African Babul Blue ■ *Azanus jesous gamra* 18–22mm

DESCRIPTION Active butterfly coming to flowers and damp spots. **SEXES** Male is bright blue on upperwings whereas female is brownish. **FLIGHT** Low; settles frequently. **DISTRIBUTION** Found in lowlands. **LARVAL FOOD PLANTS** Larval host plant is a species of *Acacia*; Babul is the Hindi name for it.

Quaker ■ *Neopithecops zalmora dharma* 16–30mm

DESCRIPTION Upperparts brown with pale patch on middle of wings, which is more pronounced in female than in male. Underparts pale with broken parallel lines of black

flecks parallel to termen edges. Termen edge looks like chain of thin rectangles with dark central spot. Pattern is distinctive. Large black spot in middle of costal edge of under hindwing. **SEXES** Similar. **FLIGHT** Weak, fairly straight flight when flying from one plant to another. **DISTRIBUTION** Scarce species seen in wet lowlands. Distribution not well known. **LARVAL FOOD PLANTS** *Glycosmis pentaphylla* [S: Dodan-pana] (family Rutaceae).

Hampson's Hedge Blue

■ *Actyolepis lilacea moorei* 28mm

DESCRIPTION Very difficult to distinguish from widespread **Common Hedge Blue** *A. puspa felderi*. Spots and crescents on underwing larger in Hampson's, especially on under forewing. **SEXES** Female's blue upperwings have thick brown edges to costal and termen edges. **FLIGHT** Low and erratic. **DISTRIBUTION** Found in central core of the island. **LARVAL FOOD PLANTS** Butterflies in this genus are known to use plants in families Fabaceae, Malpighiaceae, Mimosaceae and Sapindaceae.

Gram Blue ■ *Euchrysops cnejus cnejus* 25–30mm

DESCRIPTION Gram Blue, Lime Blue (see p. 68) and Plains Cupid (see p. 68) have two dark spots on costa of under hindwing. In Plains Cupid and Lime Blue the first of these black spots, together with three dark spots on basal area, form almost a straight line of four dark spots (towards dorsum). Gram Blue has only three spots rather than four in this 'costa-dorsum spot row'. Small Cupids (see p. 69) also have three black spots in 'costa-dorsum spot row'. To put it clearly:

4 dark spots in costa-dorsum spot row – Lime Blue, Plains Cupid.
3 dark spots in costa-dorsum spot row – Gram Blue, Small Cupid.
2 dark spots in costa-dorsum spot row – Indian Cupid.

Lime Blue can be separated from Plains Cupid by rusty-coloured splotches in discal area being discrete and not merging into a row as in Plains Cupid. **SEXES** Female's upperwings have thick brownish edges to costa and termen edges. **FLIGHT** Rapid, circling flight around bushes. **DISTRIBUTION** Widespread; most common in dry zone. **LARVAL FOOD PLANTS** Species in family Fabaceae.

Lime Blue ■ *Chilades lajus lajus* 26–27mm

DESCRIPTION Post-discal area of under forewing has an arc of irregularly shaped dark spots. See also identification notes under Gram Blue (p. 67). In Lime Blue the two black tornal spots are not fringed with rufous as in Plains Cupid (see below). **SEXES** Female's blue upperwings have thick brown edges to costa and termen edges. **FLIGHT** Low, frequently settling on flowers. **DISTRIBUTION** Throughout the island. **LARVAL FOOD PLANTS** Larval host plants include citrus species, which is unusual for a blue. Larvae are defended by red ants. Larval food plants are *Atalantia ceylanica*, *A. monophylla*, *Citrus aurantifolia* [S: Hin-dehi, Udu-dehi], *C. sinensis* [S: Peni-dodan, Punchi-jambola] and *Limonia acidissima* [S: Diwul] (family Rutaceae).

Plains Cupid ■ *Chilades pandava lanka* 25–35mm

DESCRIPTION See identification notes under Gram Blue (p. 67). **SEXES** Male and female both have blue upperwings. Male has thin brown border on termen on upper forewing. Female has thick brown border on costa and termen on upper forewing. **FLIGHT** Fast and erratic, with a few twists and turns. **DISTRIBUTION** Common to mid-hills. In some countries also referred to as Cycad Blue, in reference to it being found near *Cycad* species. Occurs even in town gardens that have its host plants. **LARVAL FOOD PLANTS** *Cycas nathorstii*, endemic to Sri Lanka, and cultivated *C. revoluta*, native to southern Japan (family Cycadaceae). At the time of writing its use of native *C. circinalis* [S: Madu], which is confined to Sri Lanka and the Andaman Islands, has not been confirmed. Larvae are attended by ants of genera *Lepisiota* and *Campanotus*. **ADULT FOOD PLANTS** I have seen it feeding on introduced Kurunegala Daisy *Tridax procumbens*.

Female

Small Cupid ■ *Chilades parrhasius nila* 22mm

DESCRIPTION Similar to Plains Cupid (see opposite), which is larger. On upper surface Small Cupid is more purplish. In under forewing double white row is formed by inwards pointing 'arrowheads'. See also identification notes under Gram Blue (p. 67). It has three not four dark spots on 'costal-dorsum row'. Both Small and Plains Cupids have ovoidal patches in discal area of under hindwing. These are fringed with a white margin. In Small Cupid ovoidal patches are slightly more discrete and less fused together than in Plains Cupid. **SEXES** Male has blue upperwings. Female's upperwings can look brown because of blue being confined to small triangle coming out of base. **DISTRIBUTION** Confined to northern dry lowlands. **LARVAL FOOD PLANTS** *Acacia leucophloea* [S: Maha Andara, Katu Andara] and *A. eburnea* (family Mimosacae).

Female

Eastern Grass Jewel ■ *Freyeria putli* 17–19mm

DESCRIPTION May be the smallest butterfly in Sri Lanka. Under hindwing has arc of eyespots on margin, which are fringed with rusty-red. **SEXES** Similar in pattern, with female being slightly larger than male. **FLIGHT** Fluttering, close to the ground. **DISTRIBUTION** Found throughout lowlands in grassland. **LARVAL FOOD PLANTS** Plants in families Fabaceae and Boraginaceae.

Ceylon Silverline ■ *Spindasis ictis ceylanica* 25–28mm

DESCRIPTION On upper forewing there is an orange post-discal patch. Dry-season form is not as richly marked on underside as wet-season form. Can be told apart from other silverlines by under forewing being relatively plain along edge of termen. Other silverlines in Sri Lanka have brick-coloured band of uneven thickness running along edge of termen. First 'string of sausages' along termen on under forewing should also be examined when differentiating silverlines. The following should be checked: thickness of brownish-red lining of sausages in relation to silver filling of sausages; how distinct the individual sausages in the string are, and where they connect with each other. **SEXES** Similar. **DISTRIBUTION** Confined to dry zone. **LARVAL FOOD PLANTS** *Dendropthoe* sp. (family Loranthaceae).

Centaur Oakblue ■ *Arhopala pseudocentaurus pirama* 53–62mm

DESCRIPTION Similar to Large Oakblue (see opposite), but much plainer on underwings. Lacks dark contrasting wavy markings on underwing of Large Oakblue, as well as dark

tornal spot. **SEXES** Female has broad black margins on upperwings, especially on upper forewing costal margin, which has a thick black margin. In male it is all metallic blue. **FLIGHT** Fast. **DISTRIBUTION** Widespread except in highest elevations. **LARVAL FOOD PLANTS** Probably similar to Large Oakblue's.

Large Oakblue ■ *Arhopala amantes amantes* 52–55mm

DESCRIPTION This species has a habit of clambering through leafy branches more like a vertebrate animal than a butterfly. I have seen it rotating like a Plum Judy (see p. 53) when it alights on leaves. Under hindwing marked with spots and wavy lines. In similar Centaur Oakblue (see opposite), under hindwing is relatively plain. Under forewing in Large also more heavily marked than in Centaur. **SEXES** In female blue upperwings are broadly margined with a dark outline. **FLIGHT** Swift. Does not fly far. **DISTRIBUTION** Widespread except in highest elevations. Found near watercourses, which may be a reflection of the presence of its host plant. In the Colombo District I have seen it in the Kotte Marshes, and one has even strayed into my garden in Colombo 08, which is less than 1km away from marshland. **LARVAL FOOD PLANTS** *Syzygium cumini* [S: Madan, Mahandan] (family Myrtaceae) and *Terminalia chebula* [S: Aralu] (family Combretaceae). Food plant in wet zone possibly a *Syzygium* or *Terminalia*. Many plants have been attributed to this species in other countries where it is found. Larvae may have symbiotic relationship with red ants *Oecophylla smaragdina*.

Aberrant Oakblue ■ *Arhopala abseus mackwoodi* 30–35mm

DESCRIPTION About half the size of more common Large and Centaur Oakblues (see above and opposite). Underwing heavily marked with dark circles outlined in white. Apical area of under forewing is silvery. Upperwings bright blue thickly edged with black. Likes to perch high, above head height, on broadleaved rainforest tree leaves. **SEXES** Similar, but female is duller than male. **FLIGHT** Very fast and hard to track. Does not engage in much circling or zigzagging. Settles quickly. Does not spend much time in the air. **DISTRIBUTION** Little known, but may be confined to good-quality lowland rainforests. This was a 'lost butterfly' until the mid-2000s, when butterfly watchers discovered that a patch close to the research station in Sinharaja was a reliable location for it. It is a canopy species like most oakblues, but will descend and perch on low-lying bushes or branches. **LARVAL FOOD PLANTS** *Vateria copallifera* (family Dipterocarpaceae).

Common Acacia Blue ■ *Surendra quercetorum discalis* 31–33mm

DESCRIPTION Male has broad black borders on costal and termen edges surrounding metallic blue upper forewings. Upper hindwing has blue completely enclosed by black

borders. Female has light brown upperwings. Upper forewings have pale patch in middle. **SEXES** Different, as described above. **DISTRIBUTION** Lowlands to mid-hills. **LARVAL FOOD PLANTS** Introduced *Calliandra surinamensis* (family Fabaceae), *Acacia caesia* [S: Heenguru Vel] and *A. pennata* [S: Hinguru, Godahinguru] (family Mimosaceae).

Redspot ■ *Zesius chrysomallus* 38–42mm

DESCRIPTION Three tails present on hindwing. Both underwings have a soft, wavy submarginal line with another band in discal area formed of rounded, hollow, rectangular shapes. Two black tornal spots capped in red. Underwing pattern is not shared with any

other blue. **SEXES** Females have two colour forms. In one, females are browner on upperwing and show little blue. In the other, they are completely brown on upperwing. **FLIGHT** Rapid. **DISTRIBUTION** Historically had a wide distribution in lowlands to 700m. At present a very scarce butterfly. I was surprised to come across one in the private 'One Acre' reserve at Talangama. **LARVAL FOOD PLANTS** *Terminalia* spp. (family Combretaceae), *Loranthus* spp. (family Loranthaceae), *Anarcardium occidentale* [S: Cadju, E: Cashewnut] (family Anacardiaceae) and plants in family Fabaceae.

Yamfly ■ *Loxura atymnus arcuata* 32mm

DESCRIPTION Orange butterfly with tail and pointed 'beak'. Not likely to be confused with any other species. **SEXES** Similar. **FLIGHT** At times seemingly weak, often settling on leaves, but I have also seen it in fast, zigzagging, low flight. **DISTRIBUTION** Found in southern half of the island, in wet zone up to 2,500m. Prefers bamboo and forested habitats. Confined to Oriental region, but not found in the Philippines. **LARVAL FOOD PLANTS** *Smilax* sp. (family Smilacaceae) and *Dioscorea* sp. (family Dioscoreaceae).

Monkey Puzzle
■ *Rathinda amor* 27mm

DESCRIPTION Distinctive butterfly. Apices of under forewing are clear brown bordered by clear white band. Rear underwing heavily marbled. No large white patches on underwing. **SEXES** Female bigger than male. **FLIGHT** Weak. **DISTRIBUTION** Distributed scarcely throughout lowlands to 700m. I have recorded it in my garden in Colombo. **LARVAL FOOD PLANTS** *Ixora* spp. (family Rubiaceae) and flowers of *Hopea* sp. (family Dipterocarpaceae).

Plane ■ *Bindahara phocides moorei* 31–32mm

DESCRIPTION Male blackish on upperwing. On hindwing termen is edged with blue. Tornal area and streamers are dirty-yellow. Upper hindwing has largish black spot with white patch against it. An arboreal butterfly seldom seen lower down. **SEXES** Female is plain brown on upperwing. White streamers. **FLIGHT** Fast flyer. **DISTRIBUTION** Historical distribution was throughout lowlands to 700m. Seems to be rare now and found in areas of good forest. **LARVAL FOOD PLANTS** *Salacia reticulata* [S: Himbutuwel, Kotala Himbutu] (family Hippocrateaeceae).

Female

Malabar Flash ■ *Rapala lankana* 35–38mm

DESCRIPTION Distinctive butterfly with orangish-brown underwing with contrasting distal line running through both underwings. Under hindwing has black tornal spot, and close to it black-and-white spot. Upperparts are brown. **SEXES** Female brighter orange-brown on underwings than male. **DISTRIBUTION** Rare butterfly confined to wet-zone forests. Endemic to Southern India and Sri Lanka. **LARVAL FOOD PLANTS** Unknown.

PIERIDAE (WHITES AND YELLOWS)

The pierids are generally coloured in shades of plain white and yellow. There are exceptions, and several species have strongly contrasting or vivid colour patterns. None of these species has 'tails' in the rear wings. All three pairs of legs are well developed and used for walking. The pierids also occur in both wet-season and dry-season forms. In the dry zone they can be seen gathered in the hundreds as they mud sip. They are also known for their large migrations in the hundreds of thousands, and two centuries ago, before Sri Lanka was heavily deforested, there may have been migrations numbering millions of butterflies.

Pioneer ▪ *Belenois aurota taprobana* 47–50mm

DESCRIPTION Distinct bold markings. Frequently mud puddles in swarms. Also visits damp spots. Seasonal dimorphism. **SEXES** Similar, with females more contrastingly marked than males. **FLIGHT** Swift and purposeful. **DISTRIBUTION** Found throughout the island but more common in dry lowlands. **LARVAL FOOD PLANTS** *Capparis sepiaria* and *Maerua arenaria* (family Capparidaceae).

Common Gull ■ *Cepora nerissa phyrne* 43–47mm

DESCRIPTION Under hindwing somewhat similar to that of Pioneer (see p. 75), but veins are not as contrasting in colour. Darker lines on veins are diffuse and not as contrastingly bold. Mud puddles in dry weather. Seasonal dimorphism. Wet-season form darker and slightly larger than dry-season form. **SEXES** Female more heavily marked on upperwing than male, especially around 'cell' on forewing, which has thick dark border. **DISTRIBUTION** Found throughout the island to 500m. Can be encountered in highlands as part of large migration. **LARVAL FOOD PLANTS** *Capparis sepiaria*, *C. roxburghii*, *C. moonii*, *C. zeylanica*, *C. brevispina* and *C. grandis* (family Capparidaceae).

Common Jezebel ■ *Delias eucharis* 65–80mm

DESCRIPTION Superficially similar to much scarcer Painted Sawtooth (see opposite), with which it may be seen. Common Jezebel differs by having reddish spots in rear edge of under hindwing, conically pointed. Painted Sawtooth has flat-edged, orangish rectangles on under hindwing's rear edge. **SEXES** Male mainly white on upperwings with veins in black. Upperwings look white in flight. Female looks dark on upperwings in flight because dark veins are broad and diffuse with many of the interspaces filled with colour. **FLIGHT** Often flies high. Rapidly fluttering wings lend the impression of a weak flyer, but in fact

it is a strong flyer that can engage in directional flight. Occasionally engages in waltzing flight without rapid, fluttering wingbeats. **DISTRIBUTION** Found throughout the island to 1,200m. Visits flowers in home gardens, roadside verges and similar places. **LARVAL FOOD PLANTS** *Dendrophthoe ligulata*, *D. falcate* and *Taxillus cuneatus* (family Loranthaceae). Parasitic loranthus plants attack soft-wooded trees most easily. The presence of trees with soft wood, such as Mango, in city gardens has enabled the spread of loranthus and in turn the Common Jezebel to exist in the hearts of crowded cities.

Painted Sawtooth ■ *Prioneris sita* 70–90mm

DESCRIPTION Superficially similar to Common Jezebel (see opposite), but Painted Sawtooth has conical tips to reddish cones on under hindwing trailing edge. **SEXES** Female heavily marked with thick dark veins. Upper forewings dark and look dark with streaks – very different from male, which looks white on upperwing. **FLIGHT** Can fly away at great speed when threatened. Otherwise adopts same leisurely flight as Common Jezebel, which it may be mimicking. The genera *Prioneris* and *Delias* are both unpalatable to birds and may engage in Müllerian mimicry (see p. 15), where they reinforce the 'message of unpalatability' to predators. **DISTRIBUTION** Found in forested areas of wet zone at 200–400m. The butterfly in the picture was mud sipping in the company of Common Jezebels in Sinharaja, which is at a lower elevation. **LARVAL FOOD PLANTS** *Capparis moonii* only has so far been confirmed in Sri Lanka, and *C. zeylanica* has been reported from India (family Capparaceae).

Striped Albatross ■ *Appias libythea libythea* 30–35mm

DESCRIPTION Both sexes show seasonal dimorphism. Wet-season form marked more strongly than dry-season form. Wet-season males have black forewing apices and black veins. Dry-season form is more white. Males mud puddle in groups. Females are absent at mud puddles and stick to flowers. Termen fairly straight compared with termens of other white pierids. **SEXES** Female marked more heavily than male in both seasonal forms, with dark edges to forewings and hindwings. **DISTRIBUTION** Found in lowlands; most common in south. **LARVAL FOOD PLANTS** *Capparis roxburghii*, *C. grandis*, *Cadaba fruticosa*, *Crateva adansonii* [S: Lunu warana] (family Capparidaceae).

Lesser Albatross ■ *Appias galene* 50–55mm ⓔ

DESCRIPTION Similar to **Common Albatross** *A. albina venusta*. Can be distinguished by forewing termen being concave rather than straight as on Common Albatross. Some authors treat this as an endemic species, *A. galene*. **SEXES** Dissimilar. Usual form is 'white form'. Male has yellow under hindwing and apical area of under forewing. On underwing of female post-discal arc in black contrasts with white apex and subapical area. On upperwing apex and subapical area is blackish with white spots. This looks like a more boldly marked pattern than that on similar upperwing of male Common Albatross. In 'yellow form' female's under hindwing and apical area of under forewing are orangish. **FLIGHT** Keeps low; fast and strong. **DISTRIBUTION** Found throughout the island. One of the key participants of large migratory swarms. Mud puddles in numbers. **LARVAL FOOD PLANTS** Butterflies in this genus use plants in family Euphobiaceae. In Sri Lanka *Drypetes sepiaria* [S: Weera] has been confirmed.

LEFT AND RIGHT: *Female*

Psyche ■ *Leptosia nina nina* 35–50mm

DESCRIPTION Easily identified in flight by white upperwings and bouncy flight low over the ground. Underwings delicately patterned in green and white. At rest underwings

camouflage it well in short grasses. **SEXES** Similar. **FLIGHT** Slow flyer, often keeping low over the ground; bouncy flight. **DISTRIBUTION** Common butterfly to about 1,000m. **LARVAL FOOD PLANTS** *Cleome gynandra* [S: Wela], *C. rutidosperma*, *Capparis roxburghii* [S: Kalu Illangedi], *C. sepiara*, *C. zeylanica* [S: Sudu Welangiriya], *Cadaba fruticosa*, *Crateva adansonii* [S: Lunu warana] (family Capparidaceae) and *Cardamine hirsute* (family Brassicaceae).

White Orange-tip ■ *Ixias marianne* 45–55mm

DESCRIPTION Combination of size and orange tip on upper surface being bordered with black helps distinguish this species from others with 'orange tips'. SEXES Dissimilar. Female has black spots on 'orange tip'. Seasonal dimorphism. Wet-season form marked more strongly than dry-season form. **FLIGHT** Low and rapid. Males settle frequently on flowers. Females fly low through undergrowth. **DISTRIBUTION** Found in dry lowlands. Most common in drier northern areas. **LARVAL FOOD PLANTS** *Capparis grandis*, *C. brevispina* and *C. sepiara* (family Capparidaceae).

Yellow Orange-tip

■ *Ixias pyrene cingalensis* 50–70mm

DESCRIPTION Very distinctive upperwings, especially upper forewing, with orange patch bordered in black around apical area. Underwings cryptically coloured and butterfly disappears when it settles in a thicket. Mud puddles occasionally. **SEXES** Female has a degree of brown mixing with yellow areas on upperwings. **FLIGHT** Fast. **DISTRIBUTION** Found in lowlands, rarely above 600m. **LARVAL FOOD PLANTS** *Capparis sepiara* has been confirmed and it has been seen ovipositing on *C. zeylanica* (family Capparidaceae).

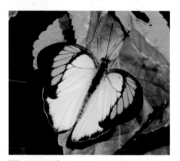

Wet-season form

Dry-season form

Dry-season form

Great Orange-tip ■ *Hebomoia glaucippe ceylonica* 80–100mm

DESCRIPTION Distinctive butterfly due to large size, and white upperwings with large red triangle on forewing apex. Red tip bordered by black. Underwings yellowish-orange and marbled. Under forewing white basally, but this does not show when butterfly is perched with wings folded. Like many orange-tip species, disappears when it lands due to cryptic camouflage. **SEXES** Male has clear white on upper hindwing. In female this is bordered by two rows of black spots. **DISTRIBUTION** Occurs throughout the island but more common at lower elevations. I have only seen it where good-quality forests still remain. **LARVAL FOOD PLANTS** *Capparis roxburghii* [S: Kalu Illangedi], *C. sepiara*, *C. moonii* and *Crateva adansonii* [S: Lunu warana] (family Capparidaceae).

Male

Small Salmon Arab ■ *Colotis amata modestus* 35–50mm

DESCRIPTION One of the few butterflies to be on the wing during the hottest times of the day. There are two forms, one white and the other salmon-pink. Large Salmon Arab (see p. 82) is similar, but termen edge of upper hindwing has neat black spots. Small Salmon Arab has diffuse dark edge to upper hindwings. **SEXES** Female has two forms. Pink form resembles male, although is slightly less bright. In 'white' form pink – especially on upperwing – is replaced by white with barely a hint of pink. **FLIGHT** Slow and low. **DISTRIBUTION** Found in dry, arid areas of lowlands. **LARVAL FOOD PLANTS** *Salvadora persica* and *Azima tetracantha* (family Salvadoraceae).

White form

Crimson Tip ■ *Colotis danae danae* 35–37mm

DESCRIPTION Confusion is possible with **Little Orange-tip** C. *etrida limbatus*, but it has a black border to 'crimson tip' on apex, and employs a slow, sailing flight. **SEXES** Dissimilar. Female has transverse row of black spots across crimson tip. Seasonal dimorphism. **FLIGHT** Fairly fast and strong. **DISTRIBUTION** Found mainly in dry northern lowlands. **LARVAL FOOD PLANTS** *Cadaba fruticosa*. In India *Capparis sepiaria* and C. *divaricata*, and *Maerua arenaria*, have been recorded. All are in family Capparidaceae.

Plain Orange-tip ■ *Colotis aurora* 35–38mm

DESCRIPTION Orange-yellow 'orange tip' thinly bordered in black on apex and small size help to identify this species. Amount of black on post-discal area bordering 'orange tip' can vary in males. **SEXES** Male is plain on underwing. **FLIGHT** Weak and low. **DISTRIBUTION** Confined to coastal areas of north and north-west. The butterfly in the picture was photographed in Mannar. **LARVAL FOOD PLANTS** *Cadaba fruticosa* (family Capparidaceae).

Female
Female

Large Salmon Arab ■ *Colotis fausta fulvia* 45–50mm

DESCRIPTION Much more brightly coloured than similar and smaller Small Salmon Arab (see p. 80). In Large Salmon Arab black margin on upper hindwing is formed by a

series of dark spots and is not a broad diffuse dark band as in Small Salmon Arab. On upper forewing dark apex does not curve back to run along costal edge as it does in Small Salmon Arab. **SEXES** Female lacks bright colours on upperwing. **FLIGHT** Can fly away fast, but may be slow when fluttering over flowers. **DISTRIBUTION** Confined to dry northern lowlands in an arc curving north of Puttalam to Trincomalee. **LARVAL FOOD PLANTS** *Maeura arenaria* (family Capparidaceae).

Male

Dark Wanderer ■ *Pareronia ceylanica ceylanica* 60–78mm

DESCRIPTION Male unlikely to be confused with any other species, as blue interspaces on upperwing are very striking. May, however, occasionally be confused with Blue Tiger (see p. 24). **SEXES** Females occur in three forms, with amount of blue decreasing and brown increasing. **FLIGHT** Male's flight hurried and undulating. Female's flight slow and undulating. **DISTRIBUTION** Found throughout the island to 500m. **LARVAL FOOD PLANTS** *Capparis zeylanica* and possibly *C. brevispinna* (family Capparidaceae).

Male

Male

Lemon Emigrant ■ *Catopsilia pomona pomona* 55–80mm

DESCRIPTION Two forms of this highly variable butterfly occur. Form *crocale* has pink-hued antennae, overall yellow colour, silver spots and red blotches on under hindwing. Form *pomona* (also referred to as form *catilla*) has black antennae, distinct black margins on forewings and is yellow overall. In form *crocale* there is yellow on basal and anal areas on creamy-white wings. Female is less yellow than male. Swarms gather at mud puddles. Underwing may or may not have red-ringed silver spots. These are also present in Mottled Emigrant (see p. 84), which has fine brownish or greenish lines, imparting a mottled effect. **SEXES** Form *crocale* female has a broader, diffuse, uneven dark border on costa and termen on upper forewing. **FLIGHT** Deep wingbeats; lightly bouncy, but generally with a sense of direction. Powerful flyer engaging in long migrations. **DISTRIBUTION** Found throughout the island. One of the species that takes part in butterfly migrations. Sometimes thousands of butterflies can be counted flying in one direction across a span of a few kilometres. **LARVAL FOOD PLANTS** *Cassia grandis*, *C. roxburghii*, *C. javanica*, *Senna alata*, *S. didymobotrya* and *S. surattenis* (family Fabaceae).

LEFT AND RIGHT: Crocale

Pomona

Pomona

Mottled Emigrant ■ *Catopsilia pyranthe pyranthe* 50–70mm

DESCRIPTION Separated from Lemon Emigrant (see p. 83) by lack of yellow in wing-bases and presence of striations on underwing. Lemon Emigrant shows yellow in wing-bases, especially on upperwing. Occasionally mud sips. One of the species that contributes to migratory swarms. **SEXES** Dissimilar. Male has finer black border on apex and smaller black cell-end spot. Underwing of female tinged yellow. **FLIGHT** Strong and jerky. **DISTRIBUTION** Found throughout the island to 500m. **LARVAL FOOD PLANTS** *Cassia grandis*, *C. fistula*, *C. auriculata*, *Senna occindentalis*, *S. tora*, *S. alata*, *S. didymobotrya* and *S. surattenis* (family Fabaceae).

IDENTIFICATION OF GRASS YELLOWS

The five species of grass yellow can be confusing in the field. The table below highlights key identification features for some of them. The spots can usually only be seen when the butterflies are examined in the hand.

	No. of spots in basal half of cell in under forewing	Termen
Three-spot Grass Yellow	3	Crenulate.
Common Grass Yellow	2	Without crenulations, almost straight.
One-spot Grass Yellow	1	Without crenulations.

Common Grass Yellow ▪ *Eurema hecabe simulata* 35–45mm

DESCRIPTION The 'poodle-head' pattern in distal area of upper forewing is clearer than in Three-spot and One-spot Grass Yellows (see p. 86). Brick-red, diffuse streak extending onto costa from subapical area sometimes present (in dry-season form) on forewing underside. However, wing edge from apex along termen remains yellow. In Three-spot a similarly coloured patch occurs, but occupies whole apical area. Species takes part in migrations. Mud sips on damp spots. **SEXES** Similar. Seasonal dimorphism. Wet-season form deeper yellow than dry-season form. **FLIGHT** Fluttering; settles frequently. Zigzag, bumbling flight. **DISTRIBUTION** Found throughout the island in grassland.
LARVAL FOOD PLANTS *Sesbania grandiflora, Pithecellobium dulce, Senna tora, Aeschynomene aspera, A. Americana, Acacia eburnea, A. leucocephala, A. nilotica* and *Albizia odoratissima* (family Fabaceae). **ADULT FOOD PLANTS** I have seen it feeding on introduced *Tridax procumbens* [E: Kurunegala Daisy].

Three-spot Grass Yellow ■ *Eurema blanda citrina* 40–50mm

DESCRIPTION Can be distinguished from very similar Common Grass Yellow (see p. 85) by a slightly crenulate forewing termen – by comparison, this is nearly straight in Common. Three spots in cell of under forewing cannot usually be seen in the field, as one or more may be covered by hindwing. 'Poodle-head' pattern" on under forewings more pronounced in Common. Three-spot has pattern that is more like a 'monkey face'. On under forewing Three-spot shows more of a 'dried blood' pattern than Common. In Three-spot it can cover a good part of the apical area, but at times the 'dried blood' pattern may not be present at all. **SEXES** Similar. **DISTRIBUTION** Common in southern half of the island, up to mid-hills. Prefers grassy habitats. **LARVAL FOOD PLANTS** Plants in family Fabaceae.

One-spot Grass Yellow ■ *Eurema ormistoni* 35–40mm

DESCRIPTION On hindwing veins end on margin with a tiny black triangle. On forewing it does not have a clear 'poodle-head' or 'monkey-face' pattern. Wings are 'rounder' than in other grass yellows. Some authors treat this as an endemic species, *Eurema ormistoni*. **SEXES** Similar. Female may be duller than male. **DISTRIBUTION** Found in wet zone to 1,000m. It is rare. **LARVAL FOOD PLANTS** *Venitlago madraspatana* and *V. gamblei* (family Rhamnaceae).

PAPILIONIDAE (SWALLOWTAILS)
These medium to large butterflies include the largest butterflies occurring in Sri Lanka. Some are powerful fliers, and a number of species are characterized by 'swallowtails'. All three pairs of legs of these butterflies are developed for walking. The birdwings found in this family include the largest butterflies in the world.

Common Bluebottle
■ *Graphium sarpedon teredon* 55–90mm

DESCRIPTION In flight confusion is possible with Common Banded Peacock (see p. 90), but at rest Common Bluebottle can be easily separated by pale contrasting blue band on its underwings and absence of 'tail'. Frequently visits flowers. **SEXES** Female slightly larger than male. Blue on females less bright. **FLIGHT** Very rapid. Tends to fly low in a fairly straight line. **DISTRIBUTION** Found throughout the island, but less common in highlands and in south. Most common in wet zone. **LARVAL FOOD PLANTS** *Cinnamomum verum* [S: Kurundu], *C. capparu-coronde*, *C. dubium*, *Neolitsea cassia* [S: Dawul kurundu] and *N. fuscata* (family Lauraceae). **ADULT FOOD PLANTS** *Lantana* spp. (family Verbenaceae).

Common Jay ■ *Graphium doson* 55–60mm

DESCRIPTION Under hindwing shows a lot of greenish-blue and has three red-fringed black spots in discal area. Underwing similar to Common Bluebottle's (see above), but in Common Jay hindwing has irregular spots in distal area, whereas the former has 'a line of thin crescents'. Red spots in hindwing of Common Bluebottle are not joined to thick pale band in middle as they are in Tailed Jay (see p. 88). They 'float' in between. The 'floaters' and 'thick crescents' are good field characteristics for distinguishing Common Bluebottle from Common Jay. **SEXES** Similar. **FLIGHT** Rapid. **DISTRIBUTION** Found up to lower hills. Absent in far north. **LARVAL FOOD PLANTS** *Polyalthia cerasoides* [S: Pattu-ul-Kenda], *P. korinti* [S: Miwenna UI kenda] and *Miliusa indica* [S: Kekili-messa] (family Annonceae).

Tailed Jay ■ *Graphium agamemnon menides* 75mm

DESCRIPTION Can be distinguished from the scarcer Common Jay (see p. 87) by lack of 'tail' on hindwing, although this feature is not always apparent. On upperwings and underwings Common Jay has fewer light spots, and it has a thicker wing-bar through the middle of the wings, tapering at both ends. Spot Swordtail (see opposite) has line of red spots on under rear wing, which is absent in Five-bar Swordtail (see opposite). I have watched Tailed Jays ovipositing on *Annona glabra*. They are very quick and place the ovipositor on the upper surface of the leaf, making contact momentarily for a second or so. They do not alight as such on the leaf and remain with their wings beating while ovipositing. **SEXES** Similar. **FLIGHT** Fast, erratic and bouncy. Sometimes I have seen this butterfly bounce up and down like a yo-yo. A hyperactive butterfly, its wings continue to beat rapidly even when it is feeding on flower nectar. **DISTRIBUTION** Found throughout

the island to 1,200m. Most common in south. **LARVAL FOOD PLANTS** *Persea americana*, *Polyalthia cerasoides* [S: Pattu-ul- Kenda], *P. suberosa* [S: Kalati], *Annona squamosa* [S: Seeni Aatha], *A. reticulata* [S: Weli Aatha], *A. glabra* [S: Wal Anoda], *A. muricata* [S: Katu-aatha], *Uvaria macropoda* [S: Attu-muddah], *U. sphenocarpa*, *Artabotrys hexapetalus* [S: Yakada-wel], *Miliusa indica* [S: Kekili-messa], *M. tomentosa* (family Annonaceae) and *Michelia champaca* (family Magnoliaceae). **ADULT FOOD PLANTS** Introduced *Lantana* spp. (family Verbenaceae).

Spot Swordtail ■ *Graphium nomius nomius* 60–65mm

DESCRIPTION Can be told apart from similar Five-bar Swordtail (see below) by upper forewing having 'spots' forming line on distal area. Spot Swordtail has black broad margin

on forewing continuing from top to bottom. In Five-bar the dark margin tapers away to form a triangle. Red band in discal area of under hindwing. **SEXES** Similar. **FLIGHT** Swift and low. Can be overlooked for a pierid because it is whitish in overall appearance on upper surface. Mud sips occasionally. **DISTRIBUTION** Found in lowlands in dry areas. **LARVAL FOOD PLANTS** *Miliusa tomenstosa* (family Annonaceae). *Polyalthia longifolia* [S: Pattu-ul-Kenda] (family Annonaceae) has been reported from India but not confirmed in Sri Lanka.

Five-bar Swordtail ■ *Pathysa antiphates ceylonicus* 70–85mm

DESCRIPTION In flight upper surface looks pale and species can be easily overlooked for one of more common whites. **SEXES** Similar. **FLIGHT** Swift and direct. **DISTRIBUTION**

Found from lowlands to lower hills. There appear to be two separate populations: one in south-west in wet lowlands (for example from Kanneliya and Sinharaja to Peak District), and another in dry lowlands of North Central Province (for example Sigiriya, Anuradhapura). **LARVAL FOOD PLANTS** *Desmos elegans* [S: Kudu-mirissa, Kukurman] and *D. zeylanica* (family Anonaceae) are possible, but unconfirmed. Dry-zone population must be utilizing one or more other plant species.

Blue Mormon ■ *Papilio polymnestor parinda* 120–150mm

IDENTIFICATION This and Common Birdwing (see p. 94) are the two largest butterflies in Sri Lanka. Male has upper hindwing in shades of blue with large, oval black spots. Upper forewing black with broad diffuse pale band. Underwing has same pattern, but black is replaced with brown and there is no blue. **SEXES** Some females can be as big as Common Birdwing. Female has two forms. One looks like male. In the other, upperwings have blue areas replaced with brown. **FLIGHT** Flies strongly with measured wingbeats. **DISTRIBUTION** Found throughout the island to 1,300m. Can be seen in home gardens as well as forests. **LARVAL FOOD PLANTS** *Citrus grandis* [S: Jambola, Jambu-naran], *C. reticulata*, [S: Hin-naran], *C. aurantifolia* [S: Hin-dehi], *C. sinenis* [S: Peni-dodan, Punchi-jambola], *Atalantia ceylanica, A. monophylla* and *Paramignya monophylla* [S: Wellangirya] (family Rutaceae).

Male

Common Banded Peacock
■ *Papilio crino* 80–100mm

DESCRIPTION Unmistakable butterfly with 'swallowtail' and iridescent green bands on upper forewings and hindwings, Both wings are crenulated. Hindwings' termen edges marked with bold eyespots. Overall impression is of an iridescent green butterfly in rapid flight. **SEXES** Similar, but female duller and browner than male. **FLIGHT** Swift and weaving. Surprisingly hard to photograph as it does not often alight close to the ground. **DISTRIBUTION** Throughout lowlands up to mid-hills. Most common in dry northern half of the island, where its host plant is most abundant. **LARVAL FOOD PLANTS** *Chloroxylon swietenia* [S: Burutha, E: Satinwood] (family Rutacea).

Red Helen ■ *Papilio helenus mooreanus* 90–150mm

DESCRIPTION Large dark butterfly with white patch on upper hindwings. Hindwing has a bold white area formed by contiguous white patches on three interspaces. Inner edge of white area is nearly straight. At rest white patch may not be visible on upperwing, but it is very evident in flight as it shows clearly on both upperwings and underwings. Occasionally mud sips. **SEXES** Female may show yellowish veins on upper forewings in post-discal area. **FLIGHT** Purposeful, directional flight. Flight pattern has a slight 'jitter'. **DISTRIBUTION** D'Abrera describes it mainly as a mountain species, over 915m. However, I hardly ever fail to see it in lowland rainforests such as Sinharaja and Morapitiya. Also found in dry and intermediate zones. Prefers forest habitats. I have seen it 'sipping' nutrients on rocks in the splash zone of fast-flowing stream in Sinharaja. **LARVAL FOOD PLANTS** *Toddalia asiatica* [S: Kudu-miris] (family Rutaceae). **ADULT FOOD PLANTS** *Clerodendron infortunatum* [S: Gas-pinna] (family Verbenaceae).

Common Mormon ■ *Papilio polytes romulus* 80–100mm

DESCRIPTION Male black with white band across discal area of upper hindwing. Forewing termen has white spots 'smudging' into edges. See 'sexes' below for notes on females. **SEXES** Females occur in three races. Race *romulus* is similar to male, with upper forewing having band of white spots on edge and upper hindwing having white band. Race *hector* resembles Crimson Rose (see p. 93), with red spots on upper hindwing and white streaks on upper forewing. Race *aristolochiae* resembles Common Rose (see p. 93), with upper rear wing having red spots and white discal area. Outer two-thirds of upper forewing have contrasting dark veins against pale background. **FLIGHT** Fast, jerky flight with rapid wingbeats. **DISTRIBUTION** Found throughout the island, but scarce above 700m. **LARVAL FOOD PLANTS** *Toddalia asiatica* [S: Kudu-miris], *Murraya koenigii* [S: Karapincha], *Citrus sinensis* [S: Peni-dodan, Punchi-jambola], *C. aurantifolia* [S: Hin-dehi, Udu-dehi], *C. limon*, *Micromelum minutum* [S: Wal-karapincha], *Atalantia ceylanica*, *Pleiospermium alatum* [S: Tumpat-Kurundu] and *Glycosmis pentaphylla* [S: Dodan-pana] (family Rutaceae).

Male

Aristolochiae *female*

Lime Butterfly

■ *Papilio demoleus demoleus* 80–100mm

DESCRIPTION Medium-sized butterfly with dark upperwings heavily marked in yellow. **SEXES** Similar. **FLIGHT** Swift. **DISTRIBUTION** Found throughout the island, becoming scarce above 500m. Common butterfly of gardens and wasteland. Common around citrus groves and where stands of Satinwood and Wood Apple are found, because these are some of the common larval food plants. **LARVAL FOOD PLANTS** *Chloroxylon swietenia* [S: Burutha E: Satinwood], *C. aurantifolia* [S: Hin-dehi, Udu-dehi], *C. limon*, *C. sinensis* [S: Peni-dodan, Punchi-jambola], *Aegle marmelos* [S: Beli], *Limonia acidissima* [S: Diwul, E: Wood Apple] (family Rutaceae), and *Cullen corylifolium* [S: Bodi] (family Fabaceae).

Common Mime ■ *Chilasa clytia lankeswara* 90–100mm

DESCRIPTION Under hindwing pattern helps to separate this species from similar tiger (Danaidae) species. Occasionally visits damp spots. **SEXES** Dissimilar and both sexes are dimorphic, having a white form and a brown form. White form mimics blue tigers and brown form imitates brown crows. Under hindwing has 'arrowheads' pointing towards wing-base and rectangular brown spots on hindwing termen margins. Hindwings are crenulate. **FLIGHT** Slow, easy flight when mimicking the tigers. Switches to fast flight when it needs to avoid danger. I once carefully observed a Common Mime that looked like a Blue Tiger (see p. 24). It used deep, rapid wingbeats, creating a fluttering impression. When covering a distance it interspersed fluttering with long glides. On the glides wings were held almost horizontal. **DISTRIBUTION** Found throughout the island to about 1,000m. Most common in southern half. **LARVAL FOOD PLANTS** *Litsea glutinosa* [S: Bomee] and *Cinnamomum verum* [S: Kurundu] (family Lauraceae). **ADULT FOOD PLANTS** Feeds on introduced *Lantana* (family Verbenaceae).

Common Rose ■ *Pachliopta aristolochiae ceylonica* 80–110mm

DESCRIPTION Mimicked by one of the forms of female Common Mormon (see p. 91). Ceylon Rose (see p. 94) has more distinct white on upper forewing. Common Rose has more pronounced 'zebra pattern' on upper forewing. See also description under Ceylon Rose. **SEXES** Similar. **FLIGHT** Slow but purposeful. Series of quick, deep wingbeats followed by brief downwards glide. **DISTRIBUTION** The most common swallowtail butterfly species. Found throughout the island to about 1,200m. Occurs in a variety of habitats, from home gardens to wasteland and forest patches. **LARVAL FOOD PLANTS** *Aristolochia indica* [S: Sapsanda, E: Indian Birthwort], *Aristolochia bracteolate* [S: Sapsanda] and *Thottea siliquosa* (family Aristolochiaceae).

First described Sri Lankan butterfly In 1758 the father of the modern system of binomial classification in biology, Carl Linneaus, described *Papilo hector*. This is now known as *Pachliopta hector* or the Crimson Rose, the first butterfly found in Sri Lanka to be described by science.

Crimson Rose ■ *Pachliopta hector* 90–110mm

DESCRIPTION Bright crimson body of this butterfly helps to identify it and separate it from mimics of female Common Mormon (see p. 91). **SEXES** Female may be duller on crimson areas of wings than male. **FLIGHT** Slow but purposeful. **DISTRIBUTION** Found throughout lowlands to about 300m. Most abundant in drier areas of north and south-east. **LARVAL FOOD PLANTS** *Aristolochia indica* [S: Sapsanda] and *A. bracteolata* [S: Bing Sapsanda] (family Aristolochiaceae).

Ceylon Rose
■ *Pachliopta jophon* 95–150mm ⓔ

DESCRIPTION On hindwing, area of cell away from base is white. Interspaces connecting to cell have white centres, but not extending to margins. Similar pattern also found in Common Rose (see p. 93), but white is not as extensive in Common, and also fewer interspaces are involved. Forewing of Common is more rounded. Forewing interspaces form flat-bottomed 'U' shapes. **SEXES** Similar. **FLIGHT** Directional, but does not come across as being powerful. **DISTRIBUTION** Confined to lower hills and mid-hills with good stands of forest. **LARVAL FOOD PLANTS** *Thottea siliquosa* (family Aristolochiaceae) may be its only larval food plant, which could result in restricted distribution.

Common Birdwing ■ *Troides darsius* 110–135mm ⓔ

DESCRIPTION One of the two largest butterflies in Sri Lanka. Dark with large yellow patches on upper hindwing. **SEXES** Female browner, lacking well-defined black areas of male. Yellow patch on hindwing is not prominent. White venations on upper forewing. Female of this species is the largest butterfly on the island. **FLIGHT** Slow but purposeful. Often seen flying high. **DISTRIBUTION** Found to 1,000m, but absent from north. **LARVAL FOOD PLANTS** *Aristolochia indica* [S: Sapsanda] (family Aristolochiaceae). It is possible that it may also use *Thottea siliquosa* (family Aristolochiaceae).

Male

Female

HESPERIIDAE (SKIPPERS)

Species of skipper found in Sri Lanka are small and thick bodied. They are generally in shades of brown, or orange and black, and in many the wings have varying degrees of white areas or spots. The antennae are noticeably broad based, with curved or hooked tips. All three pairs of legs are fully developed. Skippers fly in rapid bursts and their common names often refer to their modes of flight. Many species are crepuscular: active at dawn or dusk. The larvae of around three-quarters of the species feed on grasses (family Poaceae), palms (family Palmae) or gingers (family Zingiberaceae). Some skippers fold their wings roof-like over the abdomen. Other butterflies never do this, although they can hold their wings up when perched.

White-banded Awl ▪ *Hasora taminatus taminatus* 35–40mm

DESCRIPTION Under hindwing has broad white band separating grey inner area from brown outer area. Dark tornal spot on hindwing. Broad white band contrasting with grey basal area separates it from **Common Banded Awl** *H. chromus* – the only likely confusion species. One other skipper has this broad white band – **Branded Orange Awlet** *Bibasis oedipodea*. This has orange edging to hindwings around tornus. **DISTRIBUTION** Found throughout the island, for example in Sinharaja. **LARVAL FOOD PLANTS** *Milletia* sp. (family Fabaceae).

Common Small Flat ■ *Sarangesa dasahara albicilia* 28mm

DESCRIPTION Two sets of small silvery dots on upper forewing. Plain brown overall. Upper hindwing has silver-coloured, thin edge of 'hairs'. **SEXES** Female has mostly silvery hindwing with brown mainly in costal edge. **DISTRIBUTION** Recorded in dry zone of Sri Lanka. I have photographed this species in the Western Ghats in Thattekad in India, and I would speculate that it may be found in both the wet and dry zones in Sri Lanka.

Ceylon Snow Flat ■ *Tagiades japetus obscurus* 45–50mm

DESCRIPTION On upper hindwing area of white is restricted to diffuse band on termen and dorsum edges. Not as extensive as in similar Water Snow Flat (see opposite). Under hindwing has extensive white and is similar to Water Snow Flat's. Upper forewing has 2–3 silver dots in apical area. **SEXES** Similar. **FLIGHT** Territorial species that returns to its perch after a sortie. Rapid flight. **DISTRIBUTION** Found throughout the island except at high elevations. **LARVAL FOOD PLANTS** *Dioscorea oppositifolia* [S: Hiritala, Kitala] (family Dioscoreaceae).

Water Snow Flat ■ *Tagiades litigiosa ceylonica* 35–40mm

DESCRIPTION In male amount of white on upper hindwing is much more extensive than in Ceylon Snow Flat (see opposite). In both sexes upper hindwing has distinct spots on termen and this may be the best way to separate it from Ceylon Snow Flat. Furthermore, both sexes have 'horseshoe' of white speckles in apical area of upper forewing. **SEXES** Female has more white on upper hindwing than male. **DISTRIBUTION** Found throughout the island. **LARVAL FOOD PLANTS** *Dioscorea* spp. (family Dioscoreaceae) and *Smilax* spp. (family Smilacaceae).

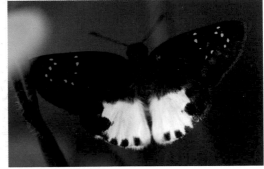

Male

Ceylon Golden Angle ■ *Caprona alida lanka* 30–35mm

DESCRIPTION Similar to **Golden Angle** *C. ransonnettii*. Golden Angle's upper forewing seems to have less white markings, although both species are variable. **SEXES** In both species females have largely white upper hindwing. In Ceylon Golden Angle female has dark spots throughout. Female Golden Angle has two dark spots on costal edge. **FLIGHT** Slow compared with quick, jerky flight of Golden Angle. **DISTRIBUTION** Confined to small area in east that includes Nilgala, where this photograph was taken. It seems very rare – between the time when this photograph was taken in Nilgala in November 2003 and the time of going to press, no one else had seen it. **LARVAL FOOD PLANTS** *Helicteres isora* [S: Liniya] (family Sterculiaceae).

Hedge Hopper ■ *Baracus vittatus* 25–35mm

DESCRIPTION Under hindwing has long pale band running through middle. This is quite distinctive and makes the species easy to identify. Upper hindwing cream coloured with borders edged with brown. **SEXES** Male has cream-coloured basal to discal area on upper hindwing; less so in female. **FLIGHT** Short bursts. **DISTRIBUTION** Found from lower hills to highlands in wet zone. Easy to see in Horton Plains National Park. **LARVAL FOOD PLANTS** Possibly grasses (family Poaceae).

Bush Hopper ■ *Ampittia dioscorides singa* 22–28mm

DESCRIPTION Has the same brownish-orange ground colour as some species of dart, but is heavily marked with brown. None of the darts are as heavily flecked with light markings.

Furthermore, some dart species have clear bands on wings, a feature absent in Bush Hopper. Termen on hindwing is rounded and there is a submarginal arc of spots on under hindwing. **SEXES** Female paler than male. **FLIGHT** Flies low, darting; not as fast as other skippers. **DISTRIBUTION** Found in lowlands to mid-hills. Often seen coming to flowers, including those of invasive *Lantana*. **LARVAL FOOD PLANTS** Rice (*Oryza sativa*) and various species of grass (family Poaceae).

Chestnut Bob ▪ *Lambrix salsala luteipalpis* 26–30mm

DESCRIPTION Overall orangish skipper. Under hindwing has three white spots in male. Largest white spot is on cell. **SEXES** Female has more white spots than male. **FLIGHT** Darting, fast flight, close to the ground. **DISTRIBUTION** Widespread on the island. **ADULT FOOD PLANTS** Visits flowers, including those of invasive *Lantana*.

Male

Common Banded Demon ▪ *Notocrypta paralysos alysia* 35–40mm

DESCRIPTION Dark skipper with prominent white band on upper and under forewing. Similar to **Restricted Demon** *N. curvifascia curvifascia*. Key feature is that white band reaches margin on under forewing. In both species band is 'bent' or 'elbowed', when they are seen with forewings held open. Restricted Demon also has several white flecks on under forewing, while Common Banded has one white spot at most. This may be the easiest distinguishing feature. **SEXES** Similar. **FLIGHT** Swift. **DISTRIBUTION** Found in wet zone to mid-hills. **LARVAL FOOD PLANTS** Plants in family Zingiberacea. I have seen it at Bodhinagala, where Spiral Ginger *Costus speciosus* is common. It is possible that this is a larval plant.

Grass Demon ■ *Udaspes folus* 35–40mm

DESCRIPTION Distinctive brown-and-white pattern on underwings. Hindwing has a rounded termen, with wing mainly pale with costal and termen having wide, irregular brown margins. When closed, pale brown of distal area on under hindwing is seen to form pattern of brownish 'steps' against white cells. **DISTRIBUTION** Found in lowlands to mid-hills.

Indian Palm Bob ■ *Suastus gremius subgrisea* 30–35mm

DESCRIPTION Light brown wings overlaid by grey scales. Under hindwing light grey-brown with variable number of black spots. Under forewing has white spots. Upper

forewing has large white spots and yellow spot near dorsum. **Ceylon Palm Bob** *S. minuta minuta* can be distinguished from this species by much larger amount of white on under hindwing. **SEXES** Female lighter than male. **FLIGHT** Low, dipping from perch to perch, hence the name Bob. **DISTRIBUTION** Widespread, but absent from highest mountains. **LARVAL FOOD PLANTS** *Calamus* spp., *Caryota urens* [S: Kithul], *Cocos cucifera* (S: Pol, E: Coconut) and *Phoenix* spp. (family Arecaceae).

Common Grass Dart ■ *Taractrocera maevius maevius* 20–25mm

DESCRIPTION Small brown skipper with white bands on abdomen. Upperwings have white flecks forming lines. **SEXES** Similar. **FLIGHT** Low and flitting. **DISTRIBUTION** Occurs to 1,220m in grassland. Scarcest in north. **LARVAL FOOD PLANTS** Grasses (family Poaceae).

Indian Dart ■ *Potanthus pallida* 22–28mm

IDENTIFICATION Easiest to identify when perched with forewings up. Open hindwings give impression of a chocolate-brown dart with orange band in middle of upper hindwing. Veins of orange band are not brown. This is a diagnostic feature, as in similar **Common Dartlet** *Oriens goloides* and **Common Dart** *Potanthus pseudomaesa pseudomaesa* the orange band has brown veins running through it. On upper hindwing Indian Dart has three diffuse orange-yellow blocks running from basal area to apex of wing at an angle to the body. **FLIGHT** Swift and low. **DISTRIBUTION** Found in wet lowlands.

Tropic Dart ■ *Potanthus satra* 22mm

Female

DESCRIPTION Upper hindwing yellow wing-band irregular, set against chocolate ground colour. In upper forewing marginal end of yellow wing-band looks as though it has been jolted out of alignment and fallen out of step. This creates a wider, dark apex. **SEXES** In male there is a wide orange-yellow costal band. Female only has a trace of this. **FLIGHT** Rapid and low, settling frequently. **DISTRIBUTION** Found in lowlands to hills. **LARVAL FOOD PLANTS** Grasses (family Poaceae).

Pale Palmdart ■ *Telicota colon amba* 30mm

DESCRIPTION Pale band on upper hindwing broken by brown venation lines, unlike in similar Tropic Dart (see above). **SEXES** Male has more orange-yellow on upper side of both forewing and hindwing. **FLIGHT** Fast and strong. **DISTRIBUTION** Found in lowlands to hills. **LARVAL FOOD PLANTS** Sugar Cane (*Saccharum officinaru*) and *Bambusa* spp. (family Poaceae).

TELLING APART THE SWIFTS				
	Little Branded *Peolopidas agna*	**Small Branded** *Pelopidas mathias*	**Large Branded** *Pelopidas subochracea*	**Conjoined** *Peolopidas conjuncta*
Underwing: white spot in hindwing cell.	Yes	Yes	Yes	Yes
Underwing: white spots on hindwing discal row.	Small spots; 3 small spots and pair of smaller white spots.	Small spots; typically 5 spots.	Large spots.	Small spots.
Upperwing ground colour.		Brown.		Rusty-brown.
Upperwing: hindwing spot near dorsum (above vein 1)				1 white spot.
Upperwing: forewing white spots on cell.		2 white spots on male. Note female has an additional white spot on dorsum.		2 white spots of equal size.
Upperwing: forewing brand present.	Yes	Yes	Yes	No

Smallest Swift ■ *Panara bada bada* 30–35mm

DESCRIPTION Can be distinguished from similar **Wallace's Swift** *Borbo cinnara* by absence of white spot in interspace 1 on upper forewing. Swifts in genus *Panara* have short antennae. **DISTRIBUTION** Found throughout lowlands. **LARVAL FOOD PLANTS** *Oryza sativa* [S: Wi, E: Rice Paddy], *Saccharum officinarum* [E: Sugar Cane], *Zea mays* [E: Maize], *Bambusa* spp. (family Poaceae) and *Colocasia esculenta* (family Araceae).

Little Branded Swift ■ *Pelopidas agna agna* 35mm

DESCRIPTION Very similar to Small Branded Swift (see opposite). Under hindwing often has five spots, with two of them being smaller than the other three and forming a pair. **FLIGHT** Short, darting flight. **DISTRIBUTION** Found in lowlands to hills. **LARVAL FOOD PLANTS** Grasses (family Poaceae).

Small Branded Swift
■ *Pelopidas mathias mathias* 35mm

DESCRIPTION Very similar to Little Branded Swift (see opposite). Small has a complete arc of spots on under hindwing. **DISTRIBUTION** Found in lower hills to highlands. Little prefers higher elevations, but both species can be found in mid-elevations. Frequently perches on flowers or on the ground. **LARVAL FOOD PLANTS** *Oryza sativa* [S: Wi, E: Rice Paddy], *Cymbopogon nardus* [S: Mana], *Saccharum officinarum* [E: Sugar Cane] and other grasses (family Poaceae).

Philippine Swift ■ *Caltoris philippina seriata* 43–44mm

DESCRIPTION A dark swift. On upper forewing three loosely rectangular white spots project backwards. Vertically above third white rectangle is a pair of loosely rectangular spots, and further up another white marking. This pattern is fairly consistent. Under hindwing white, unmarked like hindwings of other swifts, so not diagnostic. **FLIGHT** Flies higher than most skippers. May be seen on vegetation at waist height and above. **DISTRIBUTION** Found in lowland wet-zone forests with clumps of bamboo. **LARVAL FOOD PLANTS** *Bamboo* spp. (family Poaceae).

The section on classification (see p. 4) makes it clear that the modern taxonomic arrangement now groups a number of old families as subfamilies within a larger family. As an identification guide there were advantages in retaining the older arrangement, which I have done. In the new taxonomic classifications the Danaidae, Satyridae, Amathusidae, Acraeidae and Libytheidae are treated as subfamilies of the Nymphalidae. The respective subfamily treatment would be to treat them as the subfamilies Danainae, Satyrinae, Morphinae, Acraeinae and Libytheinae in the family Nymphalidae. Within families such as the Pieridae I have not split out species by the new subfamily arrangement, opting to keep the checklist deliberately simple for use by beginners.

Species in **bold** are covered in this book. E indicates an endemic species.

Danaidae (Milkweed Butterflies)

1 Glassy Tiger *Parantica aglea aglea*
2 Ceylon Tiger *Parantica taprobana*
3 Blue Glassy Tiger *Ideopsis similis exprompta*
4 Blue Tiger *Tirumala limniace exoticus*
5 Dark Blue Tiger *Tirumala septentrionis musikanos*
6 Common Tiger *Danaus genutia genutia*
7 Plain Tiger *Danaus chrysippus chrysippus*
8 Double-branded Crow *Euploea sylvester montana*
9 Great Crow *Euploea phaenareta corus*
10 Brown King Crow *Euploea klugii sinhala*
11 Common Indian Crow *Euploea core asela*
12 Ceylon Tree Nymph *Idea iasonia*

Satyridae (Satyrs or Browns)

13 Common Evening Brown *Melanitis leda leda*
14 Dark Evening Evening Brown *Melanitis phedima tambra*
15 Common Palmfly *Elymnias hypermnestra fraterna*
16 Ceylon Palmfly *Elymnias singhala*, E
17 Ceylon Treebrown *Lethe daretis*, E
18 Common Treebrown *Lethe rohria neelgheriensis*
19 Tamil Treebrown *Lethe drypetis drypetis*
20 Ceylon Forester *Lethe dynaste*, E
21 Common Bushbrown *Mycalesis perseus typhlus*
22 Dark-brand Bushbrown *Mycalesis mineus polydecta*
23 Tamil Bushbrown *Mycalesis subdita*, E
24 Cingalese Bushbrown *Mycalesis rama*, E
25 Glad-eye Bushbrown *Mycalesis patnia patnia*
26 Medus Brown *Orsotriaena medus mandata*
27 White Four-ring *Ypthima ceylonica*
28 Jewel Four-ring *Ypthima singala*, E

Amathusidae

29 Southern Duffer *Discophora lepida ceylonica*

Nymphalidae (Brush-footed Butterflies)

30 Angled Castor *Ariadne ariadne minorata*
31 Common Castor *Ariadne merione taprobana*
32 Joker *Byblia ilithyia*
33 Indian Fritillary *Argynnis hyperbius taprobana*
34 Common Leopard *Phalanta phalantha phalantha*
35 Small Leopard *Phalanta alcippe ceylonica*
36 Tamil Lacewing *Cethosia nietneri nietneri*
37 Rustic *Cupha erymanthis placida*
38 Cruiser *Vindula erota asela*
39 Tamil Yeoman *Cirrochroa thais lanka*
40 Blue Admiral *Kaniska canace haronica*
41 Indian Red Admiral *Vanessa indica nubicola*
42 Painted Lady *Vanessa cardui*
43 Chocolate Soldier *Junonia iphita iphita*
44 Yellow Pansy *Junonia hierta*
45 Blue Pansy *Junonia orithya patenas*
46 Lemon Pansy *Junonia lemonias vaisya*
47 Peacock Pansy *Junonia almana almana*
48 Grey Pansy *Junonia atlites atlites*
49 Blue Oakleaf *Kallima philarchus*
50 Autumn Leaf *Doleschallia bisaltide ceylonica*
51 Great Eggfly *Hypolimnas bolina bolina*
52 Danaid Eggfly *Hypolimnas misippus*
53 Common Sailor *Neptis hylas varmona*
54 Chestnut-streaked Sailor *Neptis jumbah nalanda*
55 Common Lascar *Pantoporia hordonia sinuata*
56 Commander *Moduza procris calidasa*
57 Clipper *Parthenos sylvia cyaneus*
58 Redspot Duke *Dophla evelina evelina*
59 Baronet *Symphaedra nais*
60 Gaudy Baron *Euthalia lubentina psittacus*
61 Baron *Euthalia aconthea vasanta*
62 Black Prince *Rohana parisatis camiba*

63 Common Nawab *Polyura athamas athamas*
64 Tawny Rajah *Charaxes psaphon psaphon*
65 Black Rajah *Charaxes solon cerynthus*

Acraeidae
66 Tawny Coster *Acraea terpsicore*

Libytheidae (The Beaks)
67 Club Beak *Libythea myrrha rama*
68 Beak *Libythea laius lepitoides*

Riodinidae
69 Plum Judy *Abisara echerius prunosa*

Lycaenidae (Blues)
70 Apefly *Spalgis epeus epeus*
71 Indian Sunbeam *Curetis thetis*
72 Dingy Lineblue *Petrelaea dana*
73 Large Four Lineblue *Nacaduba pactolus ceylonica*
74 Pale Four Lineblue *Nacaduba hermus sidoma*
75 Woodhouse's Four Lineblue *Nacaduba ollyetti*, E
76 Transparent Six Lineblue *Nacaduba kurava prominens*
77 Opaque Six Lineblue *Nacaduba beroe minima*
78 Dark Ceylon Six Lineblue *Nacaduba calauria evansi*
79 Pale Ceylon Six Lineblue *Nacaduba sinhala*, E
80 Rounded Six Lineblue *Nacaduba berenice ormistoni*
81 Common Lineblue *Prosotas nora ardates*
82 Tailless Lineblue *Prosotas dubiosa indica*
83 White-tipped Lineblue *Prosotas noreia noreia*
84 Pointed Lineblue *Ionolyce helicon viola*
85 Angled Pierrot *Caleta decidia*
86 Banded Blue Pierrot *Discolampa ethion ethion*
87 Dark Cerulean *Jamides bochus bochus*
88 Ceylon Cerulean *Jamides coruscans*, E
89 Common Cerulean *Jamides celeno tissama*
90 Milky Cerulean *Jamides lacteata*, E
91 Metallic Cerulean *Jamides alecto meilichius*
92 Forget-me-not *Catochrysops strabo strabo*
93 Silver Forget-me-not *Catochrysops panormus*
94 Pea Blue *Lampides boeticus*
95 Zebra Blue *Leptotes plinius plinius*
96 Common Pierrot *Castalius rosimon*
97 Striped Pierrot *Tarucus nara*
98 Butler's Spotted Pierrot *Tarucus callinara*
99 Dark Grass Blue *Zizeeria karsandra*
100 Lesser Grass Blue *Zizina otis indica*
101 Tiny Grass Blue *Zizula hylax hylax*
102 Indian Cupid *Everes lacturnus*
103 Red Pierrot *Talicada nyseus nyseus*
104 Bright Babul Blue *Azanus ubaldus*
105 African Babul Blue *Azanus jesous gamra*

106 Quaker *Neopithecops zalmora dharma*
107 Malayan *Megisba malaya thwaitsei*
108 Plain Hedge Blue *Celestrina lavendularis lavendularis*
109 Hampson's Hedge Blue *Actyolepis lilacea moorei*
110 Common Hedge Blue *Actyolepis puspa felderi*
111 White Hedge Blue *Udara akasa mavisa*
112 Ceylon Hedge Blue *Udara lanka*, E
113 Singalese Hedge Blue *Udara singalensis*, E
114 Gram Blue *Euchrysops cnejus cnejus*
115 Lime Blue *Chilades lajus lajus*
116 Plains Cupid *Chilades pandava lanka*
117 Small Cupid *Chilades parrhasius nila*
118 Eastern Grass Jewel *Freyeria putli*
119 Pointed Ciliate Blue *Anthene lycaenina lycaenina*
120 Common Silverline *Spindasis vulcanus fusca*
121 Plumbeous Silverline *Spindasis schistacea*
122 Clouded Silverline *Spindasis nubilis*, E
123 Ceylon Silverline *Spindasis ictis ceylanica*
124 Scarce Shot Silverline *Spindasis elima fairliei*
125 Green's Silverline *Spindasis greeni*, E
126 Long-banded Silverline *Spindasis lohita lazularia*
127 Centaur Oakblue *Arhopala pseudocentaurus pirama*
128 Large Oakblue *Arhopala amantes amantes*
129 Ormiston's Oakblue *Arhopala ormistoni*, E
130 Aberrant Oakblue *Arhopala abseus mackwoodi*
131 Tamil Oakblue *Arhopala bazaloides lanka*
132 Common Acacia Blue *Surendra quercetorum discalis*
133 Redspot *Zesius chrysomallus*
134 Purple Leaf Blue *Amblypodia anita naradoides*
135 Silver Streak Blue *Iraota timoleon nicevillei*
136 Common Tinsel *Catapaecilma major myositina*
137 Yamfly *Loxura atymnus arcuata*
138 Common Onyx *Horaga onyx cingalensis*
139 Brown Onyx *Horaga albimacula viola*
140 Monkey Puzzle *Rathinda amor*
141 Common Imperial *Cheritra freja pseudojafra*
142 White Southern Royal *Pratapa deva deva*
143 Plains Blue Royal *Tajuria jehana ceylanica*
144 Ceylon Royal *Tajuria arida*, E
145 Peacock Royal *Tajuria cippus longinus*
146 Nilgiri Tit *Hypolycaena nilgirica*
147 Cornelian *Deudorix epijarbas epijarbas*
148 Common Guava Blue *Virachola isocrates*
149 Large Guava Blue *Virachola perse ghela*
150 Plane *Bindahara phocides moorei*
151 Malabar Flash *Rapala lankana*
152 Indigo Flash *Rapala varuna lazulina*
153 Slate Flash *Rapala manea schistacea*
154 Indian Red Flash *Rapala iarbus sorya*

Pieridae (Whites and Yellows)
155 Pioneer *Belenois aurota taprobana*

156 Common Gull *Cepora nerissa phyrne*
157 Lesser Gull *Cepora nadina cingala*
158 Common Jezebel *Delias eucharis*
159 Painted Sawtooth *Prioneris sita*
160 Striped Albatross *Appias libythea libythea*
161 Chocolate Albatross *Appias lyncida taprobana*
162 Common Albatross *Appias albina venusta*
163 Lesser Albatross *Appias galene*, E
164 Plain Puffin *Appias indra narendra*
165 Psyche *Leptosia nina nina*
166 White Orange-tip *Ixias marianne*
167 Yellow Orange-tip *Ixias pyrene cingalensis*
168 Great Orange-tip *Hebomoia glaucippe ceylonica*
169 Small Salmon Arab *Colotis amata modesta*
170 Crimson Tip *Colotis danae danae*
171 Plain Orange-tip *Colotis aurora*
172 Little Orange-tip *Colotis etrida limbatus*
173 Large Salmon Arab *Colotis fausta fulvia*
174 Dark Wanderer *Pareronia ceylanica ceylanica*
175 Lemon Emigrant *Catopsilia pomona pomona*
176 Mottled Emigrant *Catopsilia pyranthe pyranthe*
177 Orange Migrant *Catopsilia scylla*
178 Small Grass Yellow *Eurema brigitta rubella*
179 Spotless Grass Yellow *Eurema laeta rama*
180 Common Grass Yellow *Eurema hecabe simulata*
181 Three-spot Grass Yellow *Eurema blanda citrina*
182 One-spot Grass Yellow *Eurema ormistoni*, E

Papilionidae (Swallowtails)
183 Common Bluebottle *Graphium sarpedon teredon*
184 Common Jay *Graphium doson doson*
185 Tailed Jay *Graphium agamemnon menides*
186 Spot Swordtail *Graphium nomius nomius*
187 Five-bar Swordtail *Pathysa antiphates ceylonicus*
188 Blue Mormon *Papilio polymnestor parinda*
189 Common Banded Peacock *Papilio crino*
190 Red Helen *Papilio helenus mooreanus*
191 Common Mormon *Papilio polytes romulus*
192 Lime Butterfly *Papilio demoleus demoleus*
193 Common Mime *Chilasa clytia lankeswara*
194 Common Rose *Pachliopta aristolochiae ceylonica*
195 Crimson Rose *Pachliopta hector*
196 Ceylon Rose *Pachliopta jophon*, E
197 Common Birdwing *Troides darsius*, E

Hesperiidae (Skippers)
198 Branded-orange Awlet *Burara oedipodea ataphus*
199 Orange-tailed Awl *Bibasis sena sena*
200 Common Banded Awl *Hasora chromus chromus*
201 White-banded Awl *Hasora taminatus taminatus*
202 Ceylon Awl *Hasora badra lanka*

203 Brown Awl *Badamia exclamationis*
204 Indian Awl King *Choaspes benjaminii benjaminii*
205 Black Flat *Celaenorrhinus spilothyrus*, E
206 Black Angle *Tapena thwaitesi thwaitesi*
207 Tricolor Pied Flat *Coladenia tissa*, E
208 Common Small Flat *Sarangesa dasahara albicilia*
209 Ceylon Snow Flat *Tagiades japetus obscurus*
210 Water Snow Flat *Tagiades litigiosa ceylonica*
211 Golden Angle *Caprona ransonnettii ransonnettii*
212 Ceylon Golden Angle *Caprona alida lanka*
213 African Marbled Skipper *Gomalia elma albofasciata*
214 Indian Skipper *Spalia galba*
215 Hedge Hopper *Baracus vittatus*, E
216 Bush Hopper *Ampittia dioscorides singa*
217 Banana Skipper *Erionota torus*
218 Decorated Ace *Thoressa decorata*, E
219 Ceylon Ace *Halpe ceylonica*, E
220 Rare Ace *Halpe egena*, E
221 Chestnut Bob *Lambrix salsala luteipalpis*
222 Common Banded Demon *Notocrypta paralysos alysia*
223 Restricted Demon *Notocrypta curvifascia curvifascia*
224 Grass Demon *Udaspes folus*
225 Indian Palm Bob *Suastus gremius subgrisea*
226 Ceylon Palm Bob *Suastus minuta minuta*
227 Tree Flitter *Hyarotis adrastus adrastus*
228 Giant Redeye *Gangara thyrsis clothilda*
229 Banded Redeye *Gangara lebadea subfasciata*
230 Common Redeye *Matapa aria*
231 Common Grass Dart *Taractrocera maevius maevius*
232 Yellow Palm Dart *Cephrenes trichopepla*
233 Common Dartlet *Oriens goloides*
234 Indian Dart *Potanthus pallida*
235 Common Dart *Potanthus pseudomaesa pseudomaesa*
236 Tropic Dart *Potanthus satra*, E
237 Pale Palmdart *Telicota colon amba*
238 Dark Palmdart *Telicota bambusae lanka*
239 Smallest Swift *Panara bada bada*
240 Wallace's Swift *Borbo cinnara*
241 Little Branded Swift *Pelopidas agna agna*
242 Large Branded Swift *Pelopidas subochracea subochracea*
243 Small Branded Swift *Pelopidas mathias mathias*
244 Conjoined Swift *Pelopidas conjuncta narooa*
245 Paintbrush Swift *Baoris penicillata*, E
246 Blank Swift *Caltoris kumara lanka*
247 Philippine Swift *Caltoris philippina seriata*

DRAGONFLIES

The species text has a focus on identification, and any behavioural aspects that assist identification. The text is succinct to fit the format of a portable photographic field guide and is aimed at the identification of the most common species, which are most likely to be encountered. More comprehensive coverage is provided in *Dragonfly Fauna of Sri Lanka* by Matjaž Bedjanič et al (2014) and the earlier and more compact edition, *Dragonflies of Sri Lanka* (2007).

STRUCTURE OF A DRAGONFLY

The overall body plan of a dragonfly mirrors that of other insects and comprises a head, thorax and abdomen.

Head The head has a relatively large pair of compound eyes with three ocelli, or simple eyes, and a pair of short antennae in an area known as the vertex. The antennae are not noticeable, unlike in butterflies and many other insects. To identify true dragonflies, it is worth noting the degree to which the eyes touch each other or do not. In damselflies the eyes are widely separated from each other and are mounted on stalks, as in a Hammerhead Shark. Behind the eyes, especially in damselflies, are coloured spots known as post-ocular spots. Below the pair of antennae is an area that juts out, the frons. In some species this is metallic in colour. Below the frons is the face and below that an area known as the labrum, below and behind which are the mandibles. As a memory aid, think of the frons as being like a nose sticking forwards above the face and the labrum being like a chin, except that the chewing parts are now below the chin.

Thorax This is the middle section of an insect's body and contains the flight muscles. It can be up to 40 per cent of the body weight in an odonate. Like all insects, dragonflies have three pairs of legs. Unusually with odonates, the six legs are forwards pointing and are not used for walking. The first pair of legs is attached to the prothorax, which joins the head to the thorax, and the middle and rear pairs are attached to the thorax. Anyone who takes up dragonfly watching soon notices that dragonflies never perch, then walk about on a plant as other insects do. They are insects that do not walk. The legs have spines that help them to catch large prey in a basket formed by their spiny legs, from which they transfer the prey to the mouth.

Wings Odonates have two pairs of wings. Again a unique feature of odonates is that the two pairs of wings can operate independently of each other. This makes odonates the most aerially manoeuvrable of all insects. In true dragonflies the forewings and hindwings have different shapes, and the hindwings are usually broader. In damselflies the wings are of the same shape, hence the insects' scientific name, Zygoptera, which means equal wings. The front vein on the wings on the leading edge is the costa, or costal vein. In the middle is a little kink or notch known as the nodus. Somewhere towards the end of the wing is a 'cell' often filled with a different colour known as the pterostigma, which means wing-spot. The venation in odonates is important for distinguishing different families and genera, but is not covered in this book because it is aimed at a non-technical readership.

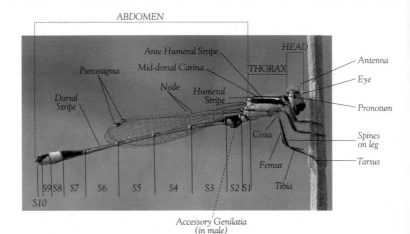

ABDOMEN

Ante Humeral Stripe

Mid-dorsal Carina

Pterostigma

Node

Humeral Stripe

HEAD

THORAX

Antenna

Eye

Pronotum

Dorsal Stripe

Coxa

Spines on leg

Femur

Tarsus

Tibia

S10 S9 S8 S7 S6 S5 S4 S3 S2 S1

Accessory Genilatia (in male)

Frons

Antenna

Vertex

Occiput

Forewing

Costa

Ante Nodal Cross Veins

Node

Pterostigma

Ovipositor (Female only)

Hindwing

Wing Base

Superior or Upper Anal Appendage (Cerci)

Superior or Upper Anal Appendages (Cerci)

Inferior or Lower Anal Appendage (only one in True Dragonflies or Anisoptera)

Occiput

Vertex

Prothorax

Eye

Frons

Clypeus

Labrum

Labium

Exoskeleton Insects do not have an internal skeleton like that of vertebrate animals such as humans. They have an external skeleton, or exoskeleton, formed of tough plates of chitinous material. The thorax is a roughly box-shaped structure. True dragonflies have a frontal plate sloping down. There may be a stripe along the top and middle of this, which is referred to as the dorsal stripe or mid-dorsal stripe. The frontal plate of the thorax joins another plate of the exoskeleton on the side of the abdomen. This join is known as the humeral suture.

There is often a pair of stripes on the frontal plate beside the mid-dorsal stripe. These are known as shoulder stripes, or ante-humeral stripes. A stripe may also be seen on the humeral suture, and is known as the humeral stripe. Sometimes there is no stripe, but the colour on the frontal plate may change at the humeral suture to a different colour on the side plates of the abdomen. The sides of the thorax may have one or more pairs of stripes, and these are referred to as side stripes or lateral stripes. The colour and thickness of these, and the ground colour (the background colour) on which they occur, are useful to note.

Abdomen This is divided into ten segments numbered 1 to 10, starting from the one nearest to the thorax. The first two are short and thick and can look as if they are part of the thorax. The last segment has what are known as terminal appendages, or anal appendages. The upper anal appendage occurs as a pair in all odonates. There is single lower one in true dragonflies and a lower pair in damselflies. The terminal appendages are also known as claspers in the males because they are used to clasp or grip females when mating. The pair of upper terminal appendages is also known as cerci, or superior appendages. The lower one or lower pair are known as an epiproct or epiprocts, respectively.

In true dragonflies, below the epiproct there is a pair of paraprocts that are small. In damselflies the paraprocts are more obvious than the pair of epiprocts. These terminal appendages are unique to each species; they are important for telling apart similar species and are used by scientists for this purpose. All damselflies and some families of true dragonfly have what is known as a complete ovipositor in females under segment 9. This is a tube-shaped structure with cutting edges that allows them to inserts their eggs into plant matter, which is usually submerged in water. In some species, under segment 9 instead of an ovipositor a subgential plate or a vulvar lamina is seen. In some odonates the males may show a narrowing of the abdomen in segments 1 and 2, forming a waist, but in the females the abdomen is always thicker and straighter.

Dragonfly Biology

The life cycle of dragonflies is often described as that of a submarine in the larval stage and a helicopter in the adult stage. In both stages dragonflies are voracious predators. The word Odonata comes from Greek, meaning 'toothed mandible'. Odonates do not have teeth and only the large true dragonflies can inflict a bite that can be felt, if they are caught. Worldwide there are around 6,000 species of odonate with more and more species being described each year.

Egg laying Odonate lives begin as an egg placed either in water (exophytic oviposition) or within plant matter (endophytic oviposition). Most damselflies and dragonflies in a few families have cutting edges on their ovipositors for making an incision in plant material in which to place eggs, which are typically elongated in shape. The majority of dragonflies dip their abdomens in water to deposit eggs that are oval in shape. A few species lay eggs in tree holes or mosses, or in ephemeral puddles created by rainwater.

Larvae On hatching, what is known as the prolarva makes its way to water. This stage is very short lived and may be only a few seconds in duration. The larva, or nymph, grows in many larval stages, or instars, undergoing numerous moults to shed its old skin as it keeps growing bigger. In tropical species the larval development may be accomplished in only a few weeks. In high latitudes the larval stage may take a few years. With a few exceptions of species whose larvae grow near water in moss or moist leaf litter, the larvae live underwater. The larvae of damselflies breathe through three fan-shaped gills that extrude from the tail. Dragonfly larvae are robustly built and lack the external gills. The final stage of a larva emerges from the water by climbing up a plant stem or an embankment. Unlike in butterflies there is no pupation stage between the larval stage and the adult stage.

Transition to adult stage Odonates undergo what is known as incomplete metamorphosis. After emergence the larval skin splits and the adult climbs out and initially hangs backwards, or at right angles to a stem in the case of damselflies, while gripping the supporting stem or surface with its feet. It then moves its head and body forwards and clambers out of the shed skin, known as an exuvia (plural exuviae). In some cases the exuviae are frequently the only record of the occurrence of a species in a locality, as the adults live a secretive life often perched out of view in tall vegetation, and only occasionally return to the water.

The newly emerged adult pumps haemolyph, the insect equivalent of blood, through the wings to stretch them out and pump out the veins, then withdraws it back into the body. It then has to wait for the wings to dry and harden. At this stage the dragonfly is very vulnerable, and some species emerge at night or dawn to avoid predation. After their wings have hardened, they fly away to safety. Many dragonfly species fly up into a tree where they wait a while longer for their wings to harden.

Newly emerged odonates have shiny wings, and are duller than adults and known as tenerals. The first few days or weeks are spent in feeding areas where the tenerals go through a process called maturation. During this stage they acquire the colours and patterns that make it possible for dragonfly watchers to identify them to species level. Immature males often look like adult females. As in the larval stage they are voracious predators and eat anything they can kill. They are either perchers that fly out from a vantage point to catch prey in midair or pluck it from vegetation, or they spend a lot of time on the wing, feeding as they fly around for long spells.

Reproduction When it is ready to mate a dragonfly flies to a breeding area, which is typically a body of water. Depending on the species it will have a particular preference, be it still water in a pond or lake, or running water in a stream or river. Males often guard territories, capturing females that fly in. Sometimes several males engage in aerial combat over a female. Male dragonflies transfer sperm from their genital pores at the end of the abdomen to accessory genitalia under the second abdominal segment, or S2. Depending on the species they may do this before capturing a female, or while in tandem either when perched or even in flight. In true dragonflies the accessory genitalia are more obvious than in damselflies, and contain visible structures called hamules. In damselflies the accessory genitalia are not prominent and there may only be a slight swelling under the second abdominal segment.

An odonate male captures a female using the legs to grab her by the thorax, then transfer to the tandem position. In this position it grips the female with its claspers (or terminal appendages) on the end of the last abdominal segment. Each species has claspers that provide for a unique fit on the female. Dragonflies grip females on the head (sometimes damage may be caused to the eyes of a female), and damselflies grip the prothorax of the female. For the transfer of sperm the female arcs her abdomen forwards so that it touches the male's secondary genitalia to form what is known as the wheel position. Some females grasp the male's abdomen with the legs. In some dragonflies, such as the skimmers, mature males develop a powdery pruinescence on the abdomen. The action of being gripped by a female's legs often results in scrape marks showing on the abdomen. The act of copulation can be swift – sometimes a fraction of a second in dragonflies.

The female may start spraying her eggs immediately, with the male mate guarding her in flight or remaining in tandem. In numerous damselfly species the male may remain in tandem for over an hour as the female oviposits under the water into plant stems.

In many odonates the male's penis is designed to extract or push away any sperm remaining from a previous mating. This is an extreme example of what zoologists call sperm competition. It explains why some species remain in tandem until the female has laid the fertilized eggs. This is contact guarding. They may also hover guard, with the male flying beside a female that is ovipositing into water. A female that has just mated and laid eggs may fly away from a breeding area until it is ready to mate again. The males, however, remain in the breeding area ready to mate with the next female. When they are in a breeding area the males may not spend much time feeding, with their time and energy focused on mating.

PHOTOGRAPHING DRAGONFLIES

Identification of many dragonfly species can be tricky. Therefore a combination of images from above that show all of the body in focus, as well as side-on shots, is very useful. For the rainforest damselflies, images are needed that show the section connecting the head to the body (the prothorax). For a few species a good image of the anal appendages is necessary to confirm the species.

Some close-up images are easiest to take with a macro lens. However, given the quality of digital images these days, even an image on a non-macro lens can sometimes be examined on a computer to provide the detail needed. This is especially so if the image is free from camera shake. It is therefore best for the camera to be placed on a tripod or other support as often as possible.

Yellow Waxtail with prey

If using a macro lens, bear in mind that the depth of field is very shallow. If you photograph from such a close distance away that the insect fills the frame, most of it may be out of focus. At times it is better to deliberately focus from a greater distance and have all of the animal in focus, rather than to get in so close that parts of it which may be important for identification go out of focus. Even if you are concentrating on a close-up of, say, just the anal appendages, it is best to get close but not too close so that you still have depth of field. An image can always be cropped later on a computer.

People who are experts on dragonflies also take images of the wing venation. Again, it is important not to get too close, and to keep the camera parallel to the plane of focus to maximize depth of field.

DRAGONFLIES AND DAMSELFLIES OF SRI LANKA

As summarized in the table below, Sri Lanka's odonate fauna comprises 129 species, of which 62 are damselflies and 67 are dragonflies. A total of 57 species is endemic, made up of 41 damselflies and 16 dragonflies. This means that 66 per cent of damselflies and 44 per cent of dragonflies are endemic. There are a further eight that are endemic at subspecies level.

Suborder and Family	No. of Species	% of All Species	No. of Endemics	% Family Endemism	No. of Endemic Subspecies
Damselflies (Zygoptera)					
Calopterygidae (jewelwings)	2	2%	1	50%	
Chlorocyphidae (jewels)	4	3%	4	100%	
Euphaeidae (gossamerwings)	1	1%	1	100%	
Lestidae (spreadwings)	6	5%	2	33%	1
Coenagrionidae (bluets)	17	14%	2	12%	2
Platycnemididae (featherlegs)	1	1%		0%	
Platystictidae (forestdamsels)	25	19%	25	100%	
Prorotoneuridae (threadtails)	6	5%	6	100%	
	62	48%	41	66%	3
Dragonflies (Anisoptera)					
Aeshnidae (hawkers)	8	6%		0%	
Gomphidae (clubtails)	14	11%	11	79%	2
Corduliidaae (emeralds)	4	3%	2	50%	2
Libellulidae (chasers)	41	33%	3	5%	1
	67	52%	16	24%	5
	129	100%	57	44%	8

THE SPECIES ACCOUNTS

The basic categories covered are description, sexes, habitat, distribution and status. Details of behaviour are sometimes included where they are not related to a general behavioural characteristic for a group. The main focus under the description is on the males, because the females of many species stay hidden. Females are discussed under sexes. The females of some genera may look very similar and more advanced texts may need to be consulted for their identification.

Scarlet Basker 'obelisking' to minimize exposure to sun

The status details are based on Bedjanič et al (2014) and use the IUCN Red List categories only for species and subspecies that are endemic (identified here with the symbol [add symbol here] after the scientific name). Details of the IUCN classification criteria are available online. Additional information on status is sometimes provided for the non-endemics under distribution. To avoid confusion, if I have added such information I have not used the IUCN Red List categories of Critically Endangered, Endangered, Vulnerable, Data Deficient, Near Threatened and Least Concern.

As far as distribution is concerned, it should be borne in mind that dragonflies are very habitat specific and that a species that is common in its aquatic habitat will probably be completely absent just a few hundred metres away.

The term dragonflies is generally used to encompass both dragonflies and damselflies. Technical jargon is avoided wherever possible, although it made sense to use established terms for important body parts, rather than to coin more familiar terms that would be misleading. In this book I have thus used terms like abdomen and thorax – with regular usage, anatomical terms become no different from using everyday terms like chest, eyes and tail.

GLOSSARY

Auricles Ear-like or rounded structures on sides of second abdominal segment of certain dragonfly families, including the clubtails (family Gomphidae) and goldenrings (family Cordulegastridae).

Bifid Two-pronged, especially with reference to terminal appendages.

Crepuscular Active at dawn and dusk.

Jizz Term popular with birders to convey a quality to a bird that helps to distinguish it from similar species. Originates from a term used by fighter pilots for General Impression of Size and Shape (GISS).

Pruinescence Some dragonflies (anisopterans) develop a powdery bluish coating on the thorax and abdomen. This can render them dark and mask colours and patterns underneath. As they mature the pruinescence increases.

SUBORDER ZYGOPTERA (DAMSELFLIES)

In damselflies the hindwing and forewing are of a similar shape and size (Zygoptera means equal wing). Damselflies are fragile-looking insects, and while resting hold their wings folded along the length of the body (except in the spreadwing family). They hunt by flying slowly, hovering over vegetation and picking off prey, unlike the more robust dragonflies, which capture prey in the air during fast flight. All damselfly females have functional ovipositors underneath abdominal segment 9. Damselflies usually make incisions in plants in which they lay their eggs. This is known as endophytic oviposition. Eggs laid in this manner are usually elongated. Some damselflies may submerge themselves partially or fully when egg laying. In some species the male remains in tandem and will assist the submerged female to fly off.

CALOPTERYGIDAE (JEWELWINGS)

These damselflies, also known as demoiselles, are characterized by broad wings and brightly coloured, metallic bodies. The wings are not on a 'stalk', and the abdomen is long and thin. The legs are noticeably long and thin compared with those of other damselflies. The insects tend to perch prominently on vegetation low over the water, or on rocks. In Sri Lanka there are no endemic species, but one of the two species occurs as an endemic subspecies.

Oriental Green-wing ▪ *Neurobasis chinensis chinensis*

DESCRIPTION Male's wings when closed are bronze on a close view in good light. Can look dark, almost black at a distance. In flight, when wings open up, they expose a beautiful, luminous, iridescent green. Pale pterostigma is the best feature distinguishing female from female Black-tipped Flashwing (see opposite). **SEXES** Female duller than male, with brownish wings. Pterostigma creamy or white, and this is useful for distinguishing female from juvenile males. **HABITAT** Occurs near fast-flowing streams in good-quality forests. This and the Black-tipped Flashwing are found in the same locations, and I have often seen both species close to each other. **DISTRIBUTION** Found mainly in wet zone from lowlands to mid-hills in forest habitats.

Male   Female

Black-tipped Flashwing ■ *Vestalis apicalis* ⓔ

DESCRIPTION Male has transparent wings with a black tip that easily distinguishes it from Oriental Green-wing (see opposite). Sometimes a group of males and females can be seen perched beside a stream, all 'flashing' their wings. I have never been sure whether these are synchronized display rituals, or a coincidence with many individuals displaying, all happening to be perched close together. Violet colour that shows up in the 'flash' is created by iridescence. At certain angles to the light, the violet iridescence does not show up at all. **SEXES** Female slightly duller and smaller than male. Lacks black wing-tips (also absent in juvenile males). **HABITAT** Occurs around fast-flowing streams in good-quality forests. **DISTRIBUTION** Found mainly in wet zone from lowlands to mid-hills in forest habitats. **STATUS** Least Concern.

Female

Male

CHLOROCYPHIDAE (JEWELS)
These damselflies have a characteristic shape, with wings extending beyond the tip of the abdomen and a 'big head'. They look like flies, and are often seen perched on overhanging vegetation or stones close to the surface of fast-running water. They are known as jewels as well as gems. All four species found in Sri Lanka are endemic. The Ebony Gem *Libellago corbeti* was only described in 2009 and is not included in this book. The male is largely all black.

Adam's Gem ■ *Libellago adami* ℮

DESCRIPTION Female has prominent white pterostigma. Male has black wing-tip that encompasses pterostigma. To distinguish males of this species from Ultima Gem (see opposite), examine the pattern of pale stripes on black 'saddle' of thorax. In Adam's Gem longer line running from front to rear is 'squeezed' by a second nearly parallel line starting midway and running below it. In Ultima Gem it looks more like a single thin line, with a dot 'falling off' it at the end. In mature male green-yellow dashes on segment 7 may be masked by black pruninescence, making it look dark tailed like Ultima Gem. **SEXES**

Female duller than male and lacks black wing-tip. **HABITAT** Male usually found perched low over fast-flowing water in ditches and other waterways. Female seen on vegetation away from water, at times perched above head height. **DISTRIBUTION** Common endemic in wet zone from lowlands to mid-hills. **STATUS** Least Concern.

Male

Female

Ultima Gem ■ *Libellago finalis* ⓔ

DESCRIPTION Very similar to Adam's Gem (see opposite); see text under that species for identification details, especially on lateral thoracic marking. I am familiar with Adam's Gem (a common endemic), so when I have seen an Ultima Gem in the field it has always struck me as different. One reason for this is that in Ultima Gem the bright green oblong patches are absent (at least in males) in abdominal segments 8–10, creating a 'black-tailed' gem. This is the biggest of the gems in Sri Lanka. **SEXES** Female similar to male and has more yellow spots on head. All its abdominal segments have yellowish patches. **HABITAT** Occurs near fast-flowing streams in quality forests. Survives around forested streams in places like Rassagala, despite heavy insecticide pollution from neighbouring tea estates. **DISTRIBUTION** Largely confined to wet zone; seems more common at elevations above 400m. **STATUS** Near Threatened.

Green's Gem ■ *Libellago greeni* ⓔ

DESCRIPTION Male has orange-red on abdomen. Green stripe on black saddle has a dot falling off at the end, as in Ultima Gem (see above). Male has a black wing-tip that encompasses the pterostigma. **SEXES** Female is green and black. It has a prominent white pterostigma. Like other gems, female does not have black wing-tip and is similar to female Ultima Gem. **HABITAT** Males usually found perched low over fast-flowing water in streams and rivulets in good-quality forests. **DISTRIBUTION** Locally common, mainly in wet zone, but also present in dry zone in southern half of the island. **STATUS** Least Concern.

Male   *Female*

Shining Gossamerwing ▪ *Euphaea splendens* ⓔ

DESCRIPTION Male is unmistakable with its greenish-blue shine when the wings are flicked open. Males often perch on rocks in fast-flowing hilly streams. Females are rarely

seen, preferring densely shaded vegetation and often perching high up in trees. Female has two thick lateral yellow stripes on thorax, thin yellow stripe on top of thorax and thin yellow line along sides of abdomen. End of abdomen is large and rounded. **SEXES** As above. **HABITAT** Occurs near rocky streams flowing through shaded forests. **DISTRIBUTION** Found mainly in wet zone up to mid-hills. Can also occur in neighbouring pockets of suitable habitat in intermediate and dry zones. **STATUS** Least Concern.

Male

Female

LESTIDAE (SPREADWINGS)
Almost all the species in this family hold out their wings in a similar way to true dragonflies (suborder Anisoptera), rather than holding them folded along the length of the body as is typical of damselflies. The wings have a thin 'stalk'. Many species have metallic colouration on the thorax. The upper terminal appendages in males are usually long and strongly curved. Two of the six species in Sri Lanka are endemic; the genus *Sinhalestes* is endemic and contains only one species.

Scalloped Spreadwing ■ *Lestes praemorsus decipiens*

DESCRIPTION Black spots on sides of thorax and 'three-pronged' scalloped pattern on dorsal surface of thorax. In mature adults, especially males, 'three prongs' may be indistinct due to pruinescence. Species are easy to tell apart, as White-tipped Spreadwing (see p. 122) has two back-to-back green 'fish hooks' on upper thorax. Male has sky-blue eyes. Terminal appendages are white and give it a 'white-tipped' look, although this name is used for the next species. **SEXES** Similar, but female paler than male and has grey eyes. **HABITAT** Occurs around open ponds with reeds. Seems to prefer perching in the middle rather than on fringing vegetation. **DISTRIBUTION** Scattered records from lowlands. The pond on the way to Martin's Lodge in Sinharaja has been a reliable location, but this may change due to increasing agricultural pressures.

White-tipped Spreadwing ■ *Lestes elatus*

DESCRIPTION Green area in middle of upper thorax is an easy identification feature. See also descriptions under Scalloped Spreadwing (p. 121). Note that Scalloped Spreadwing is also 'white tipped' because of its terminal appendages. Abdominal segments 9–10 are very pale in this species. **SEXES** Female similar to male, but paler and browner. **HABITAT** Occurs near open waterbodies with reeds. **DISTRIBUTION** Found throughout lowlands.

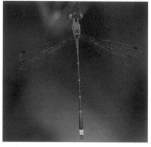

Male

Closed-wing Reedling ■ *Indolestes divisus* ⓔ

DESCRIPTION Male superficially similar to male Mountain Reedling (see opposite). In Closed-wing Reedling, black on top of second abdominal segment is more extensive and stretches to both ends of segment. Dark basal mark on segment 9 prolonged dorsally as a backwardly directed point, while in Mountain Reedling this characteristic is lacking. As the name suggests, keeps wings closed rather than 'spread-winged'. **SEXES** Female more greenish than male. **HABITAT** Occurs in reedy edges of ponds. I have seen it in a small artificial pond surrounded by vegetation. **DISTRIBUTION** Found from lowlands to highlands. Appears to be scarce. **STATUS** Endangered.

Male

Mountain Reedling ■ *Indolestes gracilis gracilis* ℮ subspecies

DESCRIPTION This spreadwing does not actually spread its wings and instead holds them closed along the axis of the body in typical damselfly manner. Male's thorax is blue on sides with a broad black stripe on top surface. Abdomen mainly dark with pale blue intersegmental rings. Segments 9–10 are blue. To distinguish this species from Closed-wing Reedling, see text under that species (opposite). **SEXES** Similarly patterned, but female much paler than male, and dark areas are replaced with brown. **HABITAT** Best seen in highlands, for example around Nuwara Eliya, in Hakgala Botanical Garden and Horton Plains National Park. **DISTRIBUTION** Found in mid-hills to highlands. Principal distribution is in central highlands. **STATUS** Vulnerable.

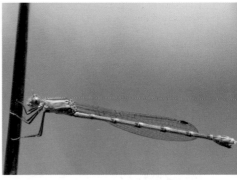

Female

Male

COENAGRIONIDAE (POND DAMSELFLIES OR BLUETS)
Worldwide, this is the most species rich family of all Odonata. Its members are adapted to a variety of habitats and some are tolerant of polluted water. Consequently, it is one of the most familiar damselfly families. Most species have clear wings that are stalked, and the females of some species have different colour forms. Species in this family tend to perch on vegetation low over the ground or over the surface of water. They inhabit and lay their eggs in static (non-flowing) water, hence the name pond damselflies. Of the 17 species recorded, only two are endemic to Sri Lanka, but there are also two endemic subspecies.

Wandering Wisp ■ *Agriocnemis pygmaea*

DESCRIPTION So small that it is easily overlooked unless you make a special effort to examine damp shorelines of waterbodies. Tends to perch on short-cropped grasses. Given the usual height of an adult person, it is almost too small to be noticed with the naked eye. Male can be distinguished from male of **White-backed Wisp** *A. femina* by the latter's very different anal appendages. Females of these two related species have different colour forms and are not easy to tell apart without consulting a field guide. **SEXES** Females have different colour forms. An isochrome form is similar to male, and another form has less black and is similar to juvenile male. There is also a red form. **HABITAT** Occurs on grassy edges by ponds and lakes. Also in ditches and canals. **DISTRIBUTION** Found throughout lowlands, ascending to mid-hills. Not rare, although as mentioned easily overlooked because of its small size and tendency to be on vegetation that hugs the ground.

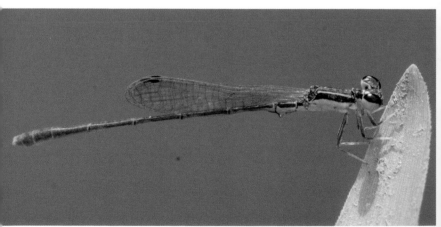

Male

Sri Lanka Midget ■ *Mortonagrion ceylonicum* ⓔ

DESCRIPTION Tiny size. Both sexes have a tear-shaped blue postocular spot, and dark abdomens. **SEXES** Female has clear blue spot on top of prothorax. Male has blackish dorsum with thin blue stripe on humeral suture stripe and thicker lateral stripe. Female has lighter blue stripes against reddish-brown dorsum. Thoracic colours in female look luminescent, and blue dorsal patches on abdominal segments are vivid. **HABITAT** Occurs around densely shaded rainforest streams. **DISTRIBUTION** Known only from a few localities in wet zone. May have been overlooked due to small size and preference for very dark, swampy areas. **STATUS** Endangered.

Male

Female

Marsh Dancer ■ *Onychargia atrocyana*

DESCRIPTION Both sexes have 'hairy faces'. Abdomen has metallic blue tinge, visible in good light. Male can be separated from male Dark-glittering Threadtail (see p. 135) by pale bluish rings at abdominal junctions. Purplish eyes. Head and thorax are black and lack purplish gloss of male Dark-glittering Threadtail. Female Marsh Dancer has yellow markings on front in space between eyes and does not have yellow spots on segments 1, 2, 8 and 9, as does female Dark-glittering Threadtail. **SEXES** Adult male's back is jet-black. Juvenile male has pair of yellow stripes. Female similar to juvenile male in having yellow stripes, but her back is brownish-black. Both sexes have thin yellow stripe joining the eyes. Female does not have pronounced ovipositor, but segments 7–10 are wider than in male. **HABITAT** Found on vegetation beside ponds and streams. Often perches at 1.2–1.5m above the ground, which makes it easy to see and photograph. **DISTRIBUTION** Found mainly in wet zone, in lowlands ascending lower hills. Common close to paddy fields and waterways. I have often found a pair in my back garden in Colombo. The nearest marsh is less than 1km away, and this may explain why occasionally one or more adults seems to be taking residence away from water for short periods. It could simply be a result of a built-in dispersal mechanism that prompts each generation to seek new breeding sites.

Male

Juvenile male

Dawn Bluetail ■ *Ischnura aurora rubilio*

DESCRIPTION Separated from male of similar Common Bluetail (see below) by not lacking black-and-blue markings on first abdominal segment (adjoining thorax). Segments 9–10 are all blue in male, which is another easy distinction from male Common Bluetail. Black on dorsal surface of abdomen confined to segments 7–8, whereas in Common Bluetail black dorsal line runs along first eight segments. **SEXES** Female similar to male but lacks blue at end of abdomen. Easily told apart from female Common Bluetail, which has black dorsal line on abdomen. **HABITAT** Occurs near ponds, lakes and other waterways. Tolerant of polluted water. **DISTRIBUTION** Found throughout the island.

Male

Common Bluetail ■ *Ischnura senegalensis*

DESCRIPTION Abdominal segments near thorax have blue on underside in both sexes, unlike in similar Dawn Bluetail (see above). Both species have greenish eyes with black on top, and bright blue postocular spots. In male Common Bluetail, segment 10 is mainly black on upper half, with blue on sides and bottom. In male Dawn Bluetail, segment 10 is blue. Male has green thorax with blackish saddle and distinct black thoracic stripe. **SEXES** Female has three colour forms, of which one is very similar to male. In one female form juvenile stage has orange thorax. **HABITAT** Occurs around ponds, lakes and other waterways. Tolerant of polluted water. **DISTRIBUTION** Found throughout the island.

Male   *Female*

Painted Waxtail
■ *Ceriagrion cerinorubellum*

DESCRIPTION Attractive and easy to identify damselfly with green eyes and thorax. Dark abdomen bright red at both ends. **SEXES** Similar. **HABITAT** Found in wide range of aquatic habitats, from still water to running water. Often encountered on vegetation-fringed footpaths close to water. **DISTRIBUTION** Largely found in south-west quadrat of the island. Given its tolerance of disturbed habitats, it is surprising that it does not have a wider distribution.

Male

Yellow Waxtail ■ *Ceriagrion coromandelianum*

DESCRIPTION Distinctive damselfly because of all-yellow abdomen. Green thorax, and top part of green eyes capped in blue. **SEXES** Similar. **HABITAT** Occurs in wide range of aquatic habitats, from marshes to ditches and open ponds fringed with vegetation. Like Painted Waxtail (see above), encountered on shrubby footpaths not far from water. Tolerant of polluted water and disturbed habitats. **DISTRIBUTION** Found throughout the island, but most common in lowlands.

Male

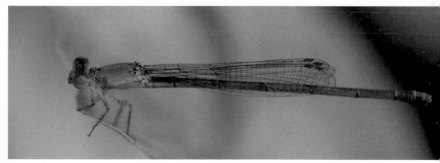

Female

Malabar Sprite ■ *Pseudagrion malabaricum*

DESCRIPTION Very similar to Blue Sprite (see opposite). The best way to distinguish them is to look at the superior terminal appendages in males. In Malabar they are not bifid and apices are curled inwards; in Blue they are bifid. Additionally, in male black areas on upper half of abdominal segments 7–9 are uniformly shaped; in Blue black areas expand towards the further end. Pterostigma is brown in Malabar and grey in Blue. If you can view the insect from above, or take a photograph from above, the following details mentioned in the older literature can be seen. Rectangular-shaped spot on segments 1–2 has a wine-glass or goblet shape with a thin stem extending (away from head) to meet a black apical ring; segments 3–7 have black dorsal markings that expand apically (away from direction of head) to meet black terminal rings; segments 8–9 are blue with black apical rings, and segment 10 has a black, nearly rectangular block. **SEXES** Female pale green overall, with thick black stripe along top of abdomen. Female Blue has thin black stripe. Female Malabar has broad black stripes on thorax that are absent in female Blue. Females of these species are

easier to distinguish in the field than males, which require closer examination. **HABITAT** Prefers still or slow-running water in habitats such as ponds, lakes and ditches. **DISTRIBUTION** Found throughout lowlands to low elevations.

Male

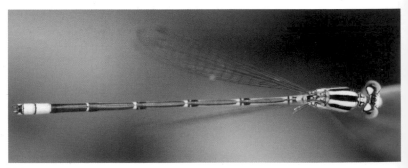

Male

Blue Sprite ■ *Pseudagrion microcephalum*

DESCRIPTION Diagnostic characteristic in male is superior terminal appendages being bifid, helping to distinguish it from similar Malabar Sprite (see opposite). Also, in segment 2 the black wine-glass or goblet shape has a thick stem that extends to black apical ring, and has an additional thin black line that extends from other side to meet a black basal ring. **SEXES** Key identification feature is thick black line on top of thorax running through middle in female Malabar, while in Blue dorsal black line is very thin. Females have green abdomen with fine black line running along top. Thorax is yellowish with pale blue dorsal stripe. **HABITAT** Occurs in wide range of habitats, including ponds, marshes, lakes and ditches with aquatic vegetation. Tolerant of polluted water and can be seen in city canals. **DISTRIBUTION** Found throughout lowlands to mid-elevations. Most abundant in lowlands.

Azure Sprite ■ *Pseudagrion decorum*

DESCRIPTION Male easy to identify as it is a pale blue damselfly with a distinctive pattern on top of thorax, with three very fine black lines. Also characteristic is the black arrowhead-shaped marking on dorsum of second abdominal segment. Top of abdomen has thick black line running along top up to segment 8. Female has similar thoracic markings, but its ground colour is green.

Although species is easy to identify there is only one record at the time of writing, from a photograph taken at Giant's Tank in Mannar in April 2006. It is common in India and more records should emerge as more observers look for it. **SEXES** See above. **HABITAT** Occurs near reed-edged ponds and lakes. **DISTRIBUTION** Found in dry lowlands.

Sri Lanka Orange-faced Sprite
■ *Pseudagrion rubriceps ceylonicum* ⓔ subspecies

DESCRIPTION Distinctive damselfly with prominent orange-red eyes contrasting with green thorax. Black line along top of abdomen finishes at segment 8, with last two abdominal segments forming 'blue tail'. **SEXES** Similar. **HABITAT** Favours stagnant pools, ditches and ponds. Prefers shaded waterbodies. **DISTRIBUTION** Found throughout lowlands. **STATUS** Least Concern.

> ## PLATYCNEMIDIDAE (FEATHERLEGS)
> Under the old taxonomic classification used in this book, Sri Lanka has only one species in this family. The featherlegs have a number of bristly spines on the legs, which have given rise to their common name. The males employ a zigzag flight, and they wave their legs in aggressive behaviour to fend off other males. According to the current taxonomy the genus *Onychargia* has been transferred from family Coenagarionidae into a new subfamily, Onychargiinae, in this family. Furthermore, family Protoneuridae has been disbanded and the two genera *Elattoneura* and *Prodasineura* have been transferred to family Platycnemididae under a new subfamily, the Disparoneurinae.

Yellow Featherleg ■ *Copera marginipes*

DESCRIPTION Yellow or orange legs are diagnostic. Clear pale 'stripe' joins tops of eyes, which have a pale band across them. Facial pattern is very striking. Juveniles are very pale, starting with white and darkening with age. White abdomen has dark spots

at segmental joints. In adults abdomen is black with last two segments being pale blue. **SEXES** Similar, but adult female is paler than male. **HABITAT** Occurs in ditches and ponds with stagnant or slow-moving water. Also seen on densely shaded footpaths close to water. **DISTRIBUTION** Found throughout the lowlands to mid-elevations. **STATUS** Very common and tolerant of disturbed habitats.

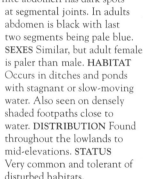

Male

Female

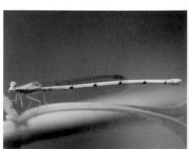

Juvenile

PLATYSTICTIDAE (FORESTDAMSELS)
This family is represented in Sri Lanka with two genera, the genus *Ceylonosticta* with 21 species and the genus *Platysticta* with 4 species. All 25 species are endemic, making Sri Lanka a biodiversity hotspot for the Platystictidae. Many of the species are similar and good photographs or examination in the hand is necessary for identification. Under a new classification, the species in the family Platystictidae found in Sri Lanka are moved into a new subfamily Platystictinae confined to Sri Lanka. A new subfamily Protostictinae now contains the other species in the rest of the Old World. Also, the genus *Ceylonosticta* has been reinstated for the exclusively endemic Sri Lankan representatives of the genus *Drepanosticta*.

Bine's Shadowdamsel ▪ *Ceylonosticta bine* ⓔ

DESCRIPTION Male has dark greenish-brown, unmarked thorax. Abdominal segments 7–9 widen out and 9–10 have contrasting pale blue on top. Abdomen dark with diffuse pale rings on inter-segmental junctions. Pronotum has prominent rounded tubercles and is hairy. Brown eyes are greenish-blue on top. SEXES Female browner than male, with pale bluish lateral stripe on thorax. Abdomen brown with diffused pale patches on top surface of inter-segmental junctions. Blue 'tail-light' of male is absent. HABITAT Occurs in good-quality rainforests, in slow-moving streams and seeps. DISTRIBUTION Found in wet zone from lowlands to higher hills. STATUS Vulnerable.

Brinck's Shadowdamsel ■ *Ceylonosticta brincki* ⓔ

DESCRIPTION Upper surface of prothorax is blue; upper surface of thorax is plain rusty-brown. Single distinct blue stripe on sides of thorax. Absence of blue stripe on

upper surface of thorax distinguishes species from **Noble Shadowdamsel** *D. digna* and **Nietner's Shadowdamsel** *D. nietneri*. **SEXES** Female similar to male except that it has larger light blue basal spots on abdominal segments 3–7. **HABITAT** Favours streams in primary and secondary rainforests. **DISTRIBUTION** Found in wet lowlands where good-quality forests remain. **STATUS** Least Concern.

Drooping Shadowdamsel ■ *Ceylonosticta lankanensis* ⓔ

DESCRIPTION Male has light bluish 'nose' and sky-blue stripe against chocolate-coloured stripe on pale brown thorax. Blue 'flash' on top at end of abdomen (segments 9–10). Upper surface of thorax has dull greenish hue. Older individuals become very dark brown, almost

black, with beautifully contrasting sky-blue markings. **SEXES** Female similar to male, but inter-segmental abdominal rings are more conspicuous. Also stouter in build. **HABITAT** Favours heavily shaded seeps and streams. Stays low and close to the ground. **DISTRIBUTION** Found in wet zone from lowlands to highlands. I have seen it in Belihul Oya in the highlands, but most records are from low elevations. **STATUS** Least Concern.

Dark Forestdamsel ■ *Platysticta apicalis* ⓔ

DESCRIPTION Male easily recognized from characteristic shape of lower anal appendages, which have a deep notch before upturned point. Dark wing-tip and completely white middle lobe of prothorax. **SEXES** Female similar to male but lacks dark wing-tip, instead having tips suffused in yellow. Black mid-dorsal stripe on sky-blue dorsum at ninth abdominal segment helps to distinguish it from the other species. **HABITAT** Frequents streams in good-quality wet-zone forests, often in densely shaded waterways. **DISTRIBUTION** Found in southern half of the island in wet zone, from lowlands to mid-elevations. **STATUS** Near Threatened.

Male

Female

Secret Forestdamsel ■ *Platysticta secreta* ⓔ

DESCRIPTION For a long time confused with **Blurry Forestdamsel** *Platysticta maculata*. Matjaž Bedjanič notes that it is easily recognized by 'the bluntly pointed bulges on the prothorax, which is coloured white and black'. May turn out to be the most common of the forest damselflies. Male has black eyes and is completely black on top of thorax. Broad pale blue lateral thoracic stripe contrasts strongly against black. Last three abdominal segments 8–10 have pale blue on top, contrasting with dark abdomen. Wings dark tipped in mature male, and suffused with yellow at tips in female. **SEXES** Female similar to male, but pattern on top of last few abdominal segments is different. **HABITAT** Occurs in rainforests. **DISTRIBUTION** Found in lowland to mid-elevation rainforests in wet zone.

PROTONEURIDAE (THREADTAILS)

In the threadtails the wings extend to around two-thirds of the length of the abdomen. The abdomen looks unusually long compared with the abdomens of other Odonata, and widens in circumference towards the tip. Sometimes the abdomen may appear to curve down; at other times it may be held straight. Many of the females in this family have hooks on the prothorax. In Sri Lanka the family is represented by two genera, *Elattoneura* with five species and *Prodasineura* with one species. All six species are endemic. In a taxonomic reclassification in 2014, family Protoneuridae was disbanded and moved into subfamily Disparoneurinae of family Platycnemididae.

Two-spotted Threadtail ▪ *Elattoneura oculata* ⓔ

DESCRIPTION Can be identified by two yellow spots on black vertex joining eyes. Brightly coloured male superficially similar to Red-striped Threadtail (see p. 136), but in

latter male has bright red eyes and female has duller red eyes. **SEXES** Female duller than male, with orange-yellow on thorax replaced by green-yellow. **HABITAT** Heavily shaded streams flowing through good-quality forests are the preferred habitat. Likes to perch low and close to water. **DISTRIBUTION** Found mainly in lowlands to mid-hills in wet zone where good-quality patches of forest remain. **STATUS** Vulnerable.

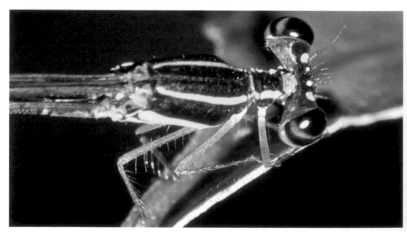

Jungle Threadtail ■ *Elattoneura caesia* ⓔ

DESCRIPTION Adult mature male easy to identify because of light blue pruinescence on upper surface of thorax, as well as light blue sides to thorax. This contrasts with thin black abdomen, and black head and eyes. There is also blue on first abdominal segment. Female lacks backwards-projecting spines on prothorax found in other *Elattoneura* females. **SEXES** Female very different from male, with thin dorsal stripe and two lateral stripes. These stripes are pale greenish-yellow. **HABITAT** Occurs around seeps and forested streams. **DISTRIBUTION** Found from mid-elevations to highlands in wet zone. **STATUS** Near Threatened.

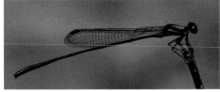

Male   *Female*

Dark-glittering Threadtail ■ *Elattoneura centralis* ⓔ

DESCRIPTION Male uniform shiny black on eyes and thorax, with metallic purple sheen. Abdomen is duller black. In dull light the whole animal looks uniform dark black. Female similar to female Jungle Threadtail (see above). Both have black saddle on thorax bisected by thin, greenish-yellow line, and thick, greenish-yellow thoracic stripe. Female of Dark-glittering Threadtail has spines on prothorax, which are a diagnostic feature. It can be seen even with the naked eye in the field if a close look is had. **SEXES** As above. **HABITAT** At sites like Bodhinagala I often find this species perched on bare rocks or on leaf litter on the ground. I once watched a female laying eggs without any mate guarding by a male. She moved her abdomen around under the water, placing eggs on submerged vegetation. **DISTRIBUTION** There are records from lowlands to mid-elevations south of very dry North Central Province. Most records are from wet zone. **STATUS** Least Concern.

Male

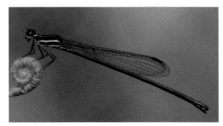

Female

Red-striped Threadtail ■ *Elattoneura tenax* ⓔ

DESCRIPTION Both sexes have red eyes. Male has bright reddish-orange dorsal stripe on thorax and another two lateral stripes. Female similarly patterned but paler. Female's red eyes avoid confusion with female Two-spotted Threadtail (see p. 134). **SEXES** As above. **HABITAT** Occurs around fast-flowing streams. **DISTRIBUTION** Found in wet zone from lowlands to about 1,800m; mostly in mid-elevations. **STATUS** Vulnerable.

Male

Male

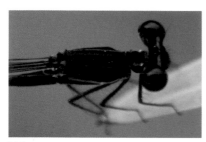

Male

Stripe-headed Threadtail
■ *Prodasineura sita* ⓔ

DESCRIPTION Common name refers to stripe on top of vertex connecting the two eyes, which also often show pale stripe. Thin dorsal stripe on thorax and pair of lateral stripes. In male 'stripe-head' is a metallic reflection and not visible unless a close view is had. Older males become very dark brown, almost black, but in contrast to Dark-glittering Threadtail (see p. 135) never have a metallic purple sheen. **SEXES** Female's head stripes and thorax paler and more conspicuous than male's. Pale stripe across frons. **HABITAT** Prefers densely vegetated waterways. Not uncommon in ditches and other waterways even in city suburbs, but easily overlooked because it is not brightly coloured or a flashy flyer. **DISTRIBUTION** Widely distributed from lowlands to lower mid-hills. Most common in wet zone. **STATUS** Least Concern.

Male

SUBORDER ANISOPTERA (DRAGONFLIES)

In dragonflies the hindwing and forewing are of different shapes and sizes (Anisoptera means unequal wing). Dragonflies are more robustly built than damselflies, and at rest tend to hold their wings out at right angles to the body and parallel to the ground. In dragonflies the eyes are close together and most often touch. Most anisopterans lay their eggs in water by flicking the abdomen in water in flight to wash off the eggs, which become attached to aquatic vegetation. This is known as exophytic oviposition, and eggs laid in this way are usually spherical or oval, and may stick together as an egg mass. Dragonfly females have functional ovipositors underneath abdominal segment 9 in the hawkers (family Aeshnidae) and petaltails (Petaluridae). The petaltails are not found in Sri Lanka.

GOMPHIDAE (CLUBTAILS)

Clubtails are a departure from the norm for true dragonflies (anisopterans) because their eyes do not touch. The degree of separation varies with species and genera, and in some the eyes are widely separated, although not at the two ends of a stalk as in damselflies (zygopterans), whose eyes are separated like those of a Hammerhead Shark. Clubtails usually have expanded segments (typically 7–9) near the tip of the abdomen, hence their common name. Males generally have bigger clubs than females. Most species have clear wings and perch horizontally on vegetation or rocks. Clubtails have 'auricles' on the sides of the second abdominal segment. Many species have strong lateral stripes, and a continuous or broken yellow line on top of the abdomen. They frequent rivers and streams as they lay their eggs in flowing water. Their hunting strategy incorporates perching – they perch low on waterside vegetation, but can also perch high up in trees, making them hard to see. Clubtails emerge as adults on horizontal and vertical surfaces. Of the 14 species found in Sri Lanka, 11 are endemic and two are species in which the Sri Lankan subspecies are endemic.

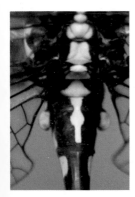

Sri Lanka Forktail showing auricles on side of second abdominal segment

Transvestite Clubtail ■ *Cyclogomphus gynostylus* ℯ

DESCRIPTION Small, overall black-and-yellow dragonfly with swollen tip of abdomen. Yellow thorax has two broad, diagonal black stripes. Male's superior terminal appendage resembles female's ovipositor turned upside down, hence the common name. **SEXES** Male has its secondary genitalia on segment 2 expanded into globular shape. This is visible on close-up views. The secondary genitalia of a freshly emerged male I photographed in October 2009 looked like a globule of transparent gel. **HABITAT** Occurs in heavily vegetated ditches and streams. Recorded in the Talangama Wetland close to the capital, Colombo. Small number of records suggests either that it occurs in naturally small populations and limited localities, or that its behaviour does not draw much attention to its presence. **DISTRIBUTION** Found in lowlands. **STATUS** Vulnerable.

Male   *Female*

Wall's Grappletail ■ *Heliogomphus walli* ℯ

DESCRIPTION The three species of *Heliogomphus* are broadly similar. Wall's Grappletail has short yellow antehumeral stripes, and two thin lateral stripes on yellow thorax. Both sexes have pale terminal appendages. In male they are lyrate and curved inwards and curled at apices. **SEXES** Similar. **HABITAT** Occurs in densely shaded, fast-flowing streams in good-quality forests. **DISTRIBUTION** Found in lowlands to mid-hills, in southern half of the island. **STATUS** Near Threatened.

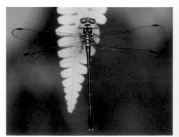

Male   *Female*

Sri Lanka Forktail ■ *Macrogomphus lankanensis* ⓔ

DESCRIPTION Superficially similar to Rivulet Tiger (see p. 140), with large size, green eyes and black body with yellow stripes. Terminal appendages are creamy-yellow in Sri Lanka Forktail and black in Rivulet Tiger. Yellow spots on middle abdominal segments are on sides in Sri Lanka Forktail and on upper surface in Rivulet Tiger, and in general yellow markings are much paler in the former species. In June 2009 I photographed an individual that stayed

perched for more than 15 minutes and allowed me to approach within 15cm of it. **SEXES** Similar. Female was not known until Amila Salgado photographed a pair mating in May 2008; they remained in wheel position for an hour and a half. Female appears to be more heavily built than male, with bolder yellow spots along abdomen. **HABITAT** Occurs around rivers, streams and irrigation channels. **DISTRIBUTION** Widespread in lowlands up to mid-hills. **STATUS** Vulnerable.

Rivulet Tiger ▪ *Gomphidia pearsoni* ℮

DESCRIPTION Large, conspicuous dragonfly, but one that is scarcely seen. Superficially similar to Rapacious Flangetail (see below), but yellow rings on abdomen are not as

extensive and black-yellow colouration is more contrasting. Pattern of two yellow stripes on black thorax is also 'less complicated' in Rivulet Tiger. **SEXES** Females not described at time of writing. **HABITAT** Favours shaded pools and slow-flowing streams in rainforests. **DISTRIBUTION** Found in wet lowlands under 900m. **STATUS** Vulnerable.

Male

Rapacious Flangetail ▪ *Ictinogomphus rapax*

DESCRIPTION Similar Rivulet Tiger (see above) has black eyes, while this species has bluish-grey eyes. Perches on reeds, bare branches and similar objects, choosing to alight on top. Tends to return to the same perch. Tolerates a close approach. Quite a distinctive species in flight because of its fast and direct flight and large size. **SEXES** Female does not have pronounced flange that male has at end of abdomen. In male this is so pronounced that it looks like the keel of a boat. **HABITAT** Occurs around small ponds and lakes with reeds or waterside vegetation, which it will perch on in the open. **DISTRIBUTION** Widespread in lowlands.

Male

Female

Aeshnidae (Hawkers)

Hawkers are robustly built dragonflies that are strong flyers. In North America they are known as darners. Their eyes meet in a broad seam on top of the head. Unlike in other dragonflies, the females have long upper terminal appendages; these often break off. The hawkers are the only true dragonfly family, or anisopterans, in Sri Lanka in which females have functional ovipositors. Unlike in most other families of anisopteran, hawkers lay their eggs on aquatic vegetation by alighting on vegetation instead of flicking the abdomen in water in flight to deposit eggs. They may at times perch vertically on a tree trunk or hang from a branch onto which they have clung. They patrol the water in a straight line unless they are zipping around catching insects. Some species can fly very high, making identification difficult. They can spend many hours on the wing, alighting only for brief periods. In the hawkers the node is near the midpoint of the costal vein. In the Libellulidae, for example, it is nearer to the apex of the wings.

Three hawker genera are found in Sri Lanka. Genus *Anax* is known as the emperors. These hawkers have a wide distribution and are strong and fast fliers. They are beautifully coloured and often patterned with blue on the thorax and abdomen. One of them, the Black Emperor, is known from only a single individual. It is believed to be a vagrant from tropical Africa. Genus *Gynacantha* is represented in Sri Lanka by one species. This is largely a tropical genus that is crepuscular. Genus *Anacaieshna* is also represented by a single species. None of the seven species of this family in Sri Lanka is endemic.

	Vagrant Emperor A. ephippiger	Pale-spotted Emperor A. guttatus	Fiery Emperor A. immaculifrons	Elephant Emperor A. indicus
	TELLING APART THE FOUR EMPERORS (MALES IN GENUS ANAX)			
Segment 2	Unmarked light blue.	Complex 'tribal-mask' pattern. Green triangle bordered below in thick black, and two more horizontal black bars that are 'bat-wing' like.	Pale blue with two black markings; triangular shape on top of a rectangle.	Green triangle bordered below with thick black.
Eyes	Brown.	Blue-green.	Blue.	Green.
Thorax	Brown.	Blue.	Blue.	Green
Terminal spots	Narrowly separated pairs on S7–9.	Widely separated pairs on S8–9.	Paired oval spots touch along abdominal mid-dorsal line, but are absent on S9–10.	Widely separated except on S10, where spots nearly touch.
Line on top of abdomen	Narrow, brown.	Broad, black.	Paired yellow spots create pattern of alternating yellow and dark rings.	Broad, dark brown.

Pale-spotted Emperor ■ *Anax guttatus*

DESCRIPTION Green eyes and unmarked green thorax are field characteristics shared with similar **Elephant Emperor** A. *indicus*. This species can be distinguished from Elephant Emperor by well-separated pale lateral spots at end of abdomen. Fiery Emperor (see opposite) has blue eyes, and sides of thorax marked strongly with blue-and-black stripes. **Vagrant Emperor** A. *ephippiger* does not have such a dark green on its thorax, and blue of abdominal segment 1 does not have shallow black 'V' and two additional parallel black bars. **SEXES** Similar. **HABITAT** Occurs near ponds and lakes. **DISTRIBUTION** Found in lowlands.

Fiery Emperor
■ *Anax immaculifrons*

DESCRIPTION Prominent blue eyes and absence of pale spots on last abdominal segment (10) help identify this species. Abdomen has fused pairs of fiery yellow spots that create impression of yellow-and-black rings on abdomen. **SEXES** In female thorax is yellowish rather than blue. **HABITAT** Favours mountain streams and pools. **DISTRIBUTION** Found from wet lowlands to highlands.

Indian Duskhawker ■ *Gynacantha dravida*

DESCRIPTION Blue eyes, brown wing venation and dark thorax and abdomen. Hints of blue where wings connect to thorax and segment 2. Abdomen pinched in at segment 3. Long terminal appendages are hairy on insides. The species is crepuscular and may be overlooked as a result. Long body, dark overall colour and emergence at dusk help identify it. **SEXES** Similar, but female has shorter terminal appendages and less pronounced pinched in waist on segment 3 than male. **HABITAT** Favours marshes; may visit garden ponds if attracted by lights. **DISTRIBUTION** Recorded mainly in wet lowlands. However, I have seen it regularly in the Kotte Marshes and Talangama Wetland, which shows that it can survive in suburban marshland.

CORDULIIDAE (EMERALDS)

Dragonflies in this family have a characteristic habit of hanging from branches and perching obliquely if they have grasped a reed. Their eyes meet in a broad seam on top of the head. Emeralds are strong flyers that spend many hours on the wing, alighting briefly. Of the four species in Sri Lanka, two are endemic and the other two are endemic subspecies. In a revised classification in 2013, genera *Epophthalmia* and *Macromia* were moved from family Corduliidae into a new family, the Macromiidae, while the treatment of genus *Macromidia* remains open.

Blue-eyed Pondcruiser ▪ *Epophthalmia vittata cyanocephala* ℮ subspecies

DESCRIPTION Distinctive blue eyes, large size, fast flight and black abdomen with golden rings make this species easy to identify. Thorax also has two golden stripes. I never get tired of being at a waterbody like Talangama Wetland and seeing a Blue-eyed Pond Cruiser fly into view. All other dragonflies are eclipsed by its arrival. It looks huge, and often flies close enough for its blue eyes to be seen with the naked eye. It then appears to disappear without trace. The species perches by 'hanging' from a tangle of branches, often above head height. As a result it is rarely seen perched, as a perch may be away from water and not easily in the field of view of an observer. This is one of the largest species on the island. **SEXES** Female similar to male, except that wings have amber tint to bases. **HABITAT** Found where waterbodies are fringed with marshes with mangrove-associate plants. **DISTRIBUTION** Widely distributed up to mid-hills where larger waterbodies are fringed with good cover. **STATUS** Near Threatened.

LIBELLULIDAE (CHASERS)

Worldwide this is the most species rich of the true dragonfly families. However, future molecular phylogenetic studies are expected to lead to numerous revisions, as many species whose family relationships are not clear are currently in this family. Members of the family vary from small dragonflies to large, showy species. Most perch conspicuously in a horizontal posture on vegetation or on the ground, or on the tips of natural or man-made objects. Their hunting strategy also involves 'perching'. They sally forth from a favourite perch to catch prey in the fashion of insect-hunting birds like bee-eaters. The eyes meet in a broad seam on top of the head. In some genera the abdomens are relatively short, and in others they are long; thus there is no typical body profile in this mixed bag of a family. At present 41 species are found in Sri Lanka, and placed in 29 genera. Just three species are endemic and one is an endemic subspecies. Except for genera *Orthetrum* (six species) and *Trithemis* (three species), all the other genera in Sri Lanka are represented by either just one or two species.

Fruhstorfer's Junglewatcher ▪ *Hylaeothemis fruhstorferi* ⓔ

DESCRIPTION Thorax has three broad yellow stripes against black background – very similar to Yerbury's Elf (see p. 146). However, Fruhstorfer's Junglewatcher's eyes are brown and green on upper and lower halves. Yerbury's Elf has eyes that are a mix of emerald green and turquoise-blue. Towards tip of abdomen, Yerbury's has incomplete yellow band. Yellow on Fruhstorfer's abdomen takes the form of longitudinal 'dashes', whereas in Yerbury's they are more like rings (bands). **SEXES** Female looks similar to male except that it has a lateral expansion on segment 8, which makes it look like a 'clubtail'. **HABITAT** Occurs near streams in large rainforests. **DISTRIBUTION** At the time of writing, this species had recently only been reliably recorded in lowland areas of Sinharaja. Look for it on the old logging road that runs from the 'barrier gate' to the research station. I have encountered it several times on the logging road, perched on vegetation about a metre above the ground, as well as near the Maguruwala by the research station. It has always tolerated a close approach for photography. The species was first collected from an elevation above Belihul Oya. **STATUS** Critically Endangered.

Female

Male

Yerbury's Elf ■ *Tetrathemis yerburii* ⓔ

DESCRIPTION Turquoise-blue eyes, small size and yellow markings against jet-black body help to distinguish this species from Fruhstorfer's Junglewatcher (see p. 145). Male Yerbury's also has dusky-tipped wing-tips compared with clear wing-tips in Fruhstorfer's.

Male

SEXES Female similar to male but lacks dusky wing-tips, which can, however, also be dark tinted at apices. **HABITAT** Yerbury's Elf was once considered very rare, but in recent years it has been recorded in good-quality lowland rainforest stretches in Sinharaja, Kithulgala and Bodhinagala, as well as in other smaller patches of forests. Refer to Fruhstorfer's Junglewatcher to avoid possible confusion. **DISTRIBUTION** Found mainly in wet lowlands to mid-hills, although there are records from highlands. Bodhinagala and Sinharaja are good sites. **STATUS** Vulnerable.

Sombre Lieutenant ■ *Brachydiplax sobrina*

DESCRIPTION Easily identified by metallic blue frons in both sexes. Male an overall greyish-blue, darkening towards tip of abdomen. Juveniles heavily marked, without bluish pruinescence, and look like females. Adult males can have trace of golden thoracic stripes. **SEXES** Female has yellowish body with black rings on abdominal segments. She looks very different to male. **Behaviour note** One day in March I saw a pair of Sombre Lieutenants mating. The male was on a perch when a female flew in. The male intercepted her in a flash and the two began mating in the air. They alighted for a few seconds on a twig, remaining joined. The female then began to lay eggs in a few places, flicking them into the water. The male mate guarded her while she was laying. She then chose a partially submerged leaf, and spent some time hovering over it. I suspected that she had now changed her strategy from shooting eggs into the water, to laying eggs perhaps one at a time by depositing them on the partially submerged leaf surface, placing each egg with her ovipositor. She then disappeared from view, as is so typical of female dragonflies. **HABITAT** Small dragonfly often found perched on reeds and rushes sticking out of water. Not easy to photograph as it tends to be in reeds away from the edges. Females are secretive and found in vegetation away from water's edge. **DISTRIBUTION** Found throughout lowlands.

Male

Female

Pale-faced Forestskimmer ■ *Cratilla lineata calverti*

DESCRIPTION Overall impression of both sexes is of blue-grey dragonfly from eyes, thorax to abdomen, with two thick yellow lateral stripes on thorax. Abdomen edged with short yellow dashes.

Labium is pale, hence the common name, but this is only evident when species is viewed head on. Pale labium contrasts with metallic frons. **SEXES** Similar. **HABITAT** Found within or close to good-quality forests near small streams. **DISTRIBUTION** Quite rare, in wet zone from lowlands to mid-hills.

Pruinosed Bloodtail ■ *Lathrecista asiatica asiatica*

DESCRIPTION Adult male easily identified by blood-red abdomen, which is shaped like a 'straight, thin red pencil'. Thorax grey in mature males; in younger males thorax is marked with lateral yellow stripes similar to female's. Frons is metallic. **SEXES** Female has red abdomen, but it is thicker than male's. Yellow thorax marked with three lateral stripes, of which two branch into 'Y' at top. **HABITAT** Prefers densely shaded undergrowth in swampy habitats or near open waterbodies. **DISTRIBUTION** Widespread in lowlands. Found in suburban wetlands but overlooked because even the males do not like perching in the open.

Male

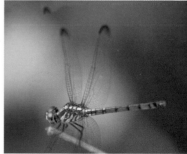

Female

Vermilion Forester ■ *Lyriothemis defonsekai* ⓔ

DESCRIPTION Male is stunning, with a broad vermilion abdomen. Both sexes have metallic frons, green-yellow thorax with prothorax-thorax join bordered in thick black, and two broad lateral black lines. First of lines branches into 'Y' at top, and the one to the rear is wide and broken in middle. Both sexes have black legs. Neither sex is likely to be confused with other species. **SEXES** In female abdomen is green-yellow with two pairs of longitudinal black lines on either side of mid-dorsal black line. **HABITAT** Known only from access path in Sinharaja, which is a disturbed edge of primary rainforest fringed and criss-crossed with streams. **DISTRIBUTION** At present only known from Sinharaja. I have encountered the female a few times on the 'logging road' used by birders for access into the Sinharaja reserve from the Kudawa entrance. **STATUS** Critically Endangered.

Male

Female

Spine-tufted Skimmer ■ *Orthetrum chrysis*

DESCRIPTION The easiest way to tell this species apart from similar looking species is by the rusty-brown thorax. This, combined with dark brown eye colour, helps eliminate many confusion species with red abdomens that have red eyes. If good views are had, red

frons can be seen. Red abdomen lacks black markings. 'Colour patch' on hindwing-base is small. Male obviously more crimson hued than male Pink Skimmer (see p. 150), which is more of a dull pink. **SEXES** Female similar to male but more orange-red than crimson. **HABITAT** Occurs in a mixture of habitats, from marshy pools to ponds and streams. **DISTRIBUTION** Found from lowlands to mid-hills. The Kudawa entrance of Sinharaja, for example, is a good site for it. Locally common.

Asian Skimmer ■ *Orthetrum glaucum*

DESCRIPTION Male similar to Marsh Skimmer (see below), but this is easily distinguished from Asian Skimmer by absence of coloured bases to hindwings. Confusion possible with Triangle Skimmer (see p. 151), but thorax is blue in Asian Skimmer, not black as in Triangle Skimmer. This also has large, diffuse black spot on bases of hindwings; in Asian Skimmer spot is amber. Triangle Skimmer additionally has a sharply defined black 'tip' at end of abdomen. Juvenile male similar to female and has a brown thorax with bold yellow lateral stripes and a blue abdomen. **SEXES** Female very different from male. Thorax is brown with two thick, pale lateral stripes. Blue-grey abdomen marked on edges with striking yellow dashes. In old females abdomen becomes pruinosed with dark blue-grey. Females have an expanded segment 8. **HABITAT** Found in damp areas, not necessarily in open or running water. Occurs on seasonally wet meadows. **DISTRIBUTION** Found from lowlands to mid-hills. Seems to prefer inland marshes to coastal ones. Common.

Juvenile male   Male

Marsh Skimmer ■ *Orthetrum luzonicum*

DESCRIPTION Male similar to male Asian Skimmer (see above). Unmarked wing-bases in Marsh Skimmer diagnostic in both sexes. **SEXES** Female very different from male. In female sides of yellow abdomen are lined with thick blackish line, and there are dark inter-segmental rings. Brown thorax has two yellow lateral lines. **Behaviour note** Male and female remain in wheel position for some time during mating. **HABITAT** Frequents marshy habitats. Sometimes seen perched on vegetation lining footpaths. **DISTRIBUTION** Found from lowlands to hills.

Female   Male

Pink Skimmer ■ *Orthetrum pruinosum neglectum*

DESCRIPTION One of the most common of 'pink-and-red' dragonflies. With experience it is possible to easily identify the male. Other males of similar species generally have much brighter crimson-red abdomens. Pink Skimmer has pink abdomen that lacks lustre. Head and thorax are dull black. **SEXES** Female has a brown and yellowish body, and is very different from male. Females are not seen often. **Behaviour note** Mating is an extended affair. The pair may fly in tandem and alight several times while coupled. Sometimes the male suspends the female, which dangles held by the head. This may be a mate-guarding strategy to prevent the female from being inseminated by another male. A pair may mate for up to half an hour. I once saw a pair engage in tandem, and the female began to lay eggs within a few minutes. The male mate guarded and chased away intruding males. **HABITAT** Occurs in all types of aquatic habitat, from paddy fields and marshes, to ditches, streams and lakes. **DISTRIBUTION** Distributed widely from lowlands to highlands.

Male

Green Skimmer ■ *Orthetrum sabina sabina*

DESCRIPTION One of the easiest ways to identify this skimmer is to look for the 'black

cylinder' it trails behind. This is formed by the last few segments (8–10) being black in colour. They are also swollen, contributing to an unusual side profile. Pale terminal appendages contrast with 'black cylinder'. First three segments also swollen, and abdomen sharply narrows for segments in middle. Overall, a greenish dragonfly with black markings. **SEXES** Similar. Terminal appendages in female are shorter, wider and appear blunt compared with pointed terminal appendages of male. **HABITAT** One of the most common dragonflies in lowlands, occurring in ponds, ditches, paddy fields and almost every waterway. **DISTRIBUTION** Found from lowlands to hills.

Triangle Skimmer ■ *Orthetrum triangulare triangulare*

DESCRIPTION Blackish head and thorax, blue pruinosed abdomen turning black on last few segments, and dark yellowish bases to hindwings. **SEXES** Female lacks blue pruinescence on abdomen. Dark thorax with two pale stripes. On abdomen segment 8 has prominent flaps on sides – their purpose is not fully understood. **HABITAT** Occurs around clear pools in montane habitats. **DISTRIBUTION** Found mainly in highlands. Horton Plains National Park is a reliable site for it.

Blue Pursuer ■ *Potamarcha congener*

DESCRIPTION Stout build lends this species a characteristic jizz. It has reddish-brown eyes, blue thorax, blue abdomen edged with golden dashes, central pair of golden lines on top of some abdominal segments, which are thinly separated, and two yellow lateral thoracic stripes. Colours of males vary with maturity, mature males being blue and having less gold. Both sexes have a habit of perching on dead exposed twigs. I have found females perching at a height of 3m or more, on trees 1km away from water. **SEXES** Female best identified by relatively laterally dilated eighth abdominal segment. It also has broad yellow stripes on thorax, helping to distinguish it from females of any similar looking species. Pale stripe along thorax extends to eyes. Abdomen has caramel-coloured edges with black stripes, which are similar to those of female Scarlet Basker (see p. 165). **HABITAT** Occurs in marshes and other open-water habitats with dense vegetation bordering them. Occasionally seen along running water such as ditches. **DISTRIBUTION** Found from lowlands to mid-hills.

Male

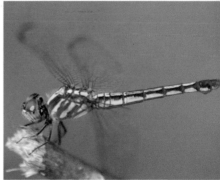

Female

Asian Pintail ■ *Acisoma panorpoides*

DESCRIPTION Distinctive profile, with swollen mid-abdominal segments, blue eyes, and thorax and first few segments of abdomen heavily marked with black against blue background. Dorsal black stripe on abdomen merges into black on last few abdominal segments. Blue eyes and pale terminal appendages. Stout appearance. **SEXES** Similar, but blue on male is replaced with brown and yellow on female. **HABITAT** Occurs near ponds and lakes that have a lot of reeds and edge vegetation. Often seen perched on leaf litter. **DISTRIBUTION** Found throughout lowlands to lower hills.

Male

Female

Orange-winged Groundling ■ *Brachythemis contaminata*

DESCRIPTION Adult male easy to identify, with an orange thorax and orange areas in middle of wing. As males mature their abdomens become more orange. Dark, two-tone line along top of abdomen. In a side view it can be seen that abdominal segments 1–5 are fattest and the segments taper down. In a top view this is not as apparent. Species has a habit of perching very close to the ground. As a result, unless it finds a twig just over the water to perch on, it is likely to be perched on dead branches on the shore. It does not like perching

on the ground or on floating vegetation. Immature male paler than adult, and has upper abdominal pattern of adult male. **SEXES** Male and female have orange and yellowish pterostigmas respectively. Female yellowish overall, with single black line along top of abdomen. **Behaviour note** This is one of the most common dragonflies when it irrupts in large numbers. I

remember a visit to a hotel beside the Wirawila Lake in April 2008. The front lawn was carpeted with Orange-winged Groundlings. I estimated there were a few hundred within a few hundred square metres. **HABITAT** Favours lakes with shoreline vegetation. **DISTRIBUTION** Found in lowlands to mid-hills.

Male

Female

Indian Rockdweller ▪ *Bradinopyga geminata*

DESCRIPTION Wings translucent with diagnostic black-and-white pterostigma. Abdomen black with jagged pale bands. Pattern on abdomen vaguely reminiscent of Green Skimmer (see p. 150). Remarkably well camouflaged against 'granite' type of rocks on which it is fond of perching. Also perches on stone paving surrounding cemented ponds. Tends to make brief flights before settling on rocky (or rock-like) substrate. **SEXES** Similar. **HABITAT** Occurs in rock pools or cement-edged ponds that simulate natural rock pools. **DISTRIBUTION** Found throughout lowlands. Uncommon, but probably often overlooked.

Oriental Scarlet ▪ *Crocothemis servilia servilia*

DESCRIPTION Male can be distinguished from male Crimson Dropwing (see p. 158) by amber (versus black) pterostigma, and forewing veins being clear (versus coloured red). From Scarlet Basker (see p. 165), it can be separated by more slender build, thin black line along upper centre of abdomen, and wing-bases not being as strongly coloured. Unlike Spine-legged Redbolt (see p. 156), it does not have contrasting paler red stripe along middle of thorax. Spine-legged Redbolt also lacks thin black line along upper centre of abdomen. This dragonfly has a habit of perching horizontally on surfaces of broad leaves, rather than clinging to edges of leaves or perching on ends of thin stems. **SEXES** Female's eyes brown on top half and bluish-grey on bottom half. Distinct black line along top centre of abdomen. Hindwings have yellowish-brown basal patch, and pterostigmas are yellowish. Similar female Crimson Dropwing has pterostigmas in dark amber that look black. **HABITAT** Occurs around ponds, marshes, paddy fields and other reed-fringed habitats. **DISTRIBUTION** Found from lowlands to mid-hills. Very common.

Male

Female

Black-tipped Percher ■ *Diplacodes nebulosa*

DESCRIPTION Male easy to identify, with its black-tipped wings, blue body and brown eyes. Females of this and Blue Percher (see below) are similar. **SEXES** Female yellow with black dorsal stripe on abdomen. Broad black line connecting thorax to prothorax, and ill-defined black thoracic stripes, help to distinguish it from similar female Blue Percher. Female Black-tipped lacks thin lateral line on segments 1–5 that is found in female Blue. Female Blue has blue eyes with reddish-brown above. Female Black-tipped has light green to slightly yellowish eyes with reddish-brown upper half. **HABITAT** Occurs around reedy lakes, ponds and marshes. **DISTRIBUTION** Found from lowlands to mid-hills.

Male

Female

Blue Percher ■ *Diplacodes trivialis*

DESCRIPTION Females of this and Black-tipped Percher (see above) are similar. Male Blue Percher has bluish eyes; male Black-tipped has chocolate-brown eyes. Upper surface of eyes of female Blue has a bluish hint; reddish upper surface in eyes of female Black-tipped. In males, amount of blue on thorax and abdomen can vary with age. One of a few species of dragonfly that habitually perch on the ground. **SEXES** Female looks like juvenile male and is more yellowish overall, with strong black markings on abdomen, including a dark, thick central line along top. Female has a more contrastingly marked abdomen, with a thin lateral line on segments 1–5, which breaks up in later segments. It looks superficially similar to female Asian Pintail (see p. 152), but the latter has a more pronouncedly swollen abdomen and spot patterning on abdominal segments 1–6. **HABITAT** Occurs in a variety of freshwater habitats; also beside brackish water. **DISTRIBUTION** Found from lowlands to mid-hills. Very common.

Male

Female

Light-tipped Demon ■ *Indothemis carnatica*

DESCRIPTION One of the smaller skimmers, dark overall with pale terminal appendages. Confusion with Indigo Dropwing (see p. 159) can be avoided, as the latter does not have pale appendages. Also, male Light-tipped Demon can have pale halves to tibia (upper part of leg). Furthermore, Indigo has prominent dark basal patch on hindwing. In Light-tipped this is a small amber spot and may not be noticeable in some individuals. Also, in Light-tipped, along top of blue abdomen is a pair of diffuse black lines that are thinly separated. Indigo does not have this feature, although Restless Demon (see below) has a similar pattern. Variable amount of yellow on abdomen depending on level of blue pruinescence. **SEXES** Female yellow overall on face, throax and abdomen, with deeper yellow on abdomen. Thick black mid-dorsal stripe on abdomen, with irregular black lateral markings on segments 1–4. Pterostigma is yellow and contrasts with pale wings. Terminal appendages black. **HABITAT** Favours lakes and ponds. **DISTRIBUTION** Found in lowlands, mainly in dry zone. Scarce species with few records.

Restless Demon ■ *Indothemis limbata sita*

DESCRIPTION Similar to Light-tipped Demon (see above) but has dark appendages and prominent dark bases to hindwings. Eyes almost black; reddish in Light-tipped. Distinct yellow dashes on segments 6–8. Distinct yellow dashes on sides of abdomen as in Indigo Dropwing (see p. 159), but black bases to hindwings are twice or three times bigger than in Indigo. **SEXES** Female black and yellow on abdomen and thorax. Amber wing-bases on hindwings. Off-white pterostigma. **HABITAT** Occurs around reedy lakes and ponds. Keeps away from edges. **DISTRIBUTION** Found in lowlands, mainly in dry zone.

Pied Parasol ■ *Neurothemis tullia tullia*

DESCRIPTION One of 3–4 of the most common and most widespread species. Can turn up in all sorts of places, sometimes a few hundred metres from the nearest water. It is not unusual to come across a cluster showing a variety of 'plumages'. Males have deep black on basal half of wings bordered with white, and transparent outer-wings. Mature males have powder-blue abdomen, sometimes looking greyish-blue. Juvenile males have yellow dorsal stripe on blue abdomen, but have dark wing-tips that distinguish them from females. **SEXES** Female yellowy-brown. Dark wing-tips diagnostic. **HABITAT** Occurs in marshes, ditches, paddy fields, lakes and similar places. A number of individuals can often be seen perched together in banks of vegetation. **DISTRIBUTION** Found from lowlands to mid-hills. Very common.

Male

Female

Spine-legged Redbolt ■ *Rhodothemis rufa*

DESCRIPTION Adults have a red abdomen. Adult retains stripe on thorax, which shows up as red stripe contrasting with browner thorax. Red stripe across thorax is a continuation of same red of abdomen. In adult eyes are red with a hint of blue on top. They meet at a point and do not meet in a broad seam, as do those of similar species. Pterostigmas are pale orange. 'Red stripe' is the best diagnostic feature distinguishing this species from similar red dragonflies. If good views are had, 'spines' on legs may be seen. Most species of Odonata have these transverse spines on their legs, but they are not always easily visible to the naked eye. Bases of hindwings marked with yellowish-orange, with just a trace of basal markings on forewings. Juveniles have very prominent yellow stripe along orange abdomen, which makes them easy to identify. **SEXES** Female similar to juveniles. **HABITAT** Common dragonfly of ponds and marshes. Likes to perch on top of a large leaf or twig, about 30cm or so above the ground, close to water. At times it may be encountered a few hundred metres away from water. **DISTRIBUTION** Found throughout the lowlands.

Male

Juvenile

Red-veined Darter ■ *Sympetrum fonscolombii*

DESCRIPTION Adult male has black markings on top of segments 8–9, and on sides of abdominal segments. This is diagnostic among red dragonflies found in Sri Lanka. Veins at bases of wings also strongly coloured in red. Fond of perching low over the ground, often choosing to perch on bare soil. **SEXES** In female red is replaced with yellow. Basal markings on wings are amber, and there is no red on abdomen. Frons is yellow in female, red in male.

Behaviour note Several males may patrol a single pond and intercept a female that enters the air space over the pond. The males may clash to mate. The first male to couple with the female will then fly off in tandem and land on some vegetation, usually close to the ground. After mating the male remains in tandem as the female flies over the water depositing her eggs. As in many dragonfly species, the male will detach himself and fly close to the ovipositing female, and mate guard her against other males. By staying in tandem, a male creates a further barrier to another male, rendering his mating void. **HABITAT** Found in temperate countries. Horton Plains National Park is one of the best sites in Sri Lanka. **DISTRIBUTION** In Sri Lanka confined to highlands.

Crimson Dropwing ■ *Trithemis aurora*

DESCRIPTION Confusion of male with Orange-winged Groundling (see p. 152) may seem unlikely, but occasional pale male may cause confusion. Females have a superficial similarity.

Male

Female

Juvenile male

Both sexes of Crimson Dropwing and Orange-winged Groundling can always be distinguished from each other by the following features. Both sexes of Crimson Dropwing have black or blackish pterostigmas. Male Orange-winged Groundling's pterostigmas are orangish; they are yellowish in female. Wing-points of Orange-winged Groundling are clear; Crimson Dropwing has dusky wing-points. Normal male Crimson Dropwing is unlikely to be confused with male Orange-winged Groundling. Typical adult male has entirely crimson body. Male's forewing veins are distinctly scarlet. Wing-bases are dark. Similar Oriental Scarlet (see p. 153) has orangish, not black pterostigma. Oriental Scarlet also has black line running along upper-surface midline of abdomen. Leading veins of forewing are not scarlet in Oriental Scarlet. In male Crimson Dropwing segment 9 is laterally edged black and segment 10 is transversely edged black on top. Segments 1 and 2 are also transversely edged black. These black markings are present on juvenile male, which is orange-yellow on abdomen. Juvenile male does not have lateral black markings on abdomen that are present in female. **SEXES** Female's body overall yellow and amber. Pterostigmas are blackish. Double line (can be a diffused blackish line at times) along centre of upper surface of abdomen, and pair of black lines along sides of abdomen. Sides of eyes light blue, and top reddish-brown. Abdomen sides straighter and parallel; in adults and juvenile male they are curved into a slightly oval shape. Immature males similar to females but lack black stripes along edges of abdomen. Upper surfaces of segments 8 and 9 black on sides and trailing transverse bar, creating orange triangular shapes within top surfaces of these segments. Yellow basal patches on hindwings not black centred as in Scarlet Basker (see p. 165). Female Crimson Dropwing has black femur and yellow tibia in legs. **HABITAT** Occurs in a variety of freshwater habitats, from slow-moving streams to ponds and lakes. Common where reed-fringed edges are available. **DISTRIBUTION** Common in Sri Lanka, from lowlands to mid-hills.

Indigo Dropwing ■ *Trithemis festiva*

DESCRIPTION Male Indigo Dropwing likely to be confused with male Light-tipped Demon (see p. 155). Latter has yellowish terminal appendages and small (can be overlooked) amber patches at bases of hindwings. By contrast, male Indigo Dropwing has clear black patch at bases of hindwings. Both have blue bodies, but in Indigo Dropwing body is brighter blue with pairs of yellow lines parallel to longitudinal axis of abdomen. Number of yellow pairs visible can be variable. Indigo Dropwing's golden dashes are in middle of upper side of abdomen. This helps prevent confusion with male Blue Pursuer (see p. 151), whose yellow dashes are on edge of abdomen. Male Indigo Dropwing has uniform blue thorax and lacks yellowish thoracic stripes of Blue Pursuer. **SEXES** Female has orange base to hindwing, blue eyes with reddish-brown upper half, and yellowish thorax with lateral brown stripes. Femur (upper half of leg) may be pruinosed greyish-blue. Abdomen yellow and strongly marked with dorsal stripe and pair of thick lateral stripes along abdomen. These are intersected by transverse bars on abdominal segments, creating pair of longitudinal rectangular shapes on each segment, except in segments 8–10, which are all dark. **HABITAT** Males often seen perched on rocks in streams flowing through forested areas. May also be found in standing water. Females rarely seen. **DISTRIBUTION** Found from mid-hills to highlands. Locally common.

Male

Female

Dancing Dropwing
■ *Trithemis pallidinervis*

DESCRIPTION Habit of perching with wings held up gives this species jizz of a long-winged dragonfly. Legs are distinctly long in genus *Trithemis*. Wings suffused with orangish-yellow. Bases a more pronounced orange-yellow. Thorax yellow with pair of lateral thoracic stripes and pair of stripes at front. Abdomen black with yellow-green inter-segmental spaces on top surface. Eyes creamy-blue with reddish-brown on top. Pterostigma dark amber bordered with pale yellow, creating two-toned effect. Terminal appendages mainly creamy-yellow. First few segments of abdomen smaller in circumference than others. Although the description seems quite wordy, in the field it is not difficult to identify the species as the combination of features gives it a specific jizz. **SEXES** Similar. Abdomen shorter and thicker in female than in male, as is the case in some dragonfly species. Terminal appendages yellow in female, black in male. **HABITAT** Lakes and ponds. Sometimes seen perched on bushes away from water. **DISTRIBUTION** Found throughout lowlands.

Female

Sapphire Flutterer ■ *Rhyothemis triangularis*

DESCRIPTION This fabulous dragonfly can look like a short-winged dark dragonfly in flight. Clear tips to wings are rendered invisible at a distance, and only blue basal halves

Male

are visible. **SEXES** Similar. **HABITAT** I have found it in muddy ditches and streams polluted with agricultural run-off and silt, but where good-quality forest is close to hand. Also found in ponds in good-quality forest patches. Rather surprisingly, on a few occasions I have also found it in the Talangama Wetland close to Colombo. **DISTRIBUTION** Most records are from south-west of the island, where lowland rainforests are found, but there are records from dry zone as well. Rare.

Variegated Flutterer ■ *Rhyothemis variegata variegata*

DESCRIPTION One of the most common dragonflies, although this does not detract from its attractiveness. Also one of the larger dragonflies, and easily noticed as it likes to perch conspicuously in the open. Wings amber and dark brown with an unmarked lead-grey abdomen. **SEXES** Male has blackish tips to wings. Female's wing-tips are clear. This is the easiest of the differences in wing patterns to look out for. In general, female more heavily marked on both wings than male. **HABITAT** Fairly pollution resistant, being found in polluted canals. Occurs in mixture of wetland habitats, from ponds, ditches and canals, to lakes. **DISTRIBUTION** Found from lowlands to mid-hills. Very common.

Female

Male

Wandering Glider ■ *Pantala flavescens*

DESCRIPTION Both from above and below, abdomen shows strong barring. Male has bluish-grey eyes crowned with rusty-red. Both sexes have long terminal appendages. **SEXES** Eyes of female olive-brown above. Otherwise generally similar to male, although lighter in colour. **Behaviour note** Swarms of these dragonflies are often seen hovering overhead, sometimes well above the heights of roofs. This is a very aerial dragonfly, seemingly spending hours on the wing without perching. It perches high, sometimes choosing telephone wires. Early in the morning I have seen females attempting to oviposit on the shiny roofs of cars. Engages in a multi-generational migration between Asia and Africa. The triangular wing with a broad base allows it to glide, which is believed to help with its long migratory flights. **HABITAT** Occurs in variety of freshwater habitats, from paddy fields to marshes and lakes. **DISTRIBUTION** Can be encountered anywhere on the island, but seems to be most prevalent in lowlands. Very common.

Burmeister's Glider ■ *Tramea basilaris burmeisteri*

DESCRIPTION Similar to Sociable Glider (see below), as in both species males have red abdomens with black rings and black patches on top of last three abdominal segments (8–10). Easiest way to distinguish both sexes of this species from Sociable Glider is to look at pattern on base of hindwing. In Burmeister's Glider it is two toned with two irregular

black patches bordered in gold. In Sociable Glider it is a contiguous dark patch without a gold border. The insects fly quite high and may be overlooked if they are perched high. **SEXES** Female's abdomen yellow with black stripes, forming 'tiger' pattern. Red in male's abdomen replaced with yellow in female. **HABITAT** Favours lakes and marshland in lowlands. **DISTRIBUTION** Found throughout lowlands.

Female

Sociable Glider ■ *Tramea limbata*

DESCRIPTION Black patch at base of hindwing uniformly dark. In similar Burmeister's Glider (see above) hindwing patch is variable and made up of two dark patches with a lighter toned patch of colour in between. Males of both species have thin red abdomens with blackish inter-segmental rings and long, pointed terminal appendages. **SEXES** Similar. **HABITAT** Favours clear, open stretches of water and visits swimming pools. Found around large lakes and ponds. **DISTRIBUTION** Widespread in lowlands. In 2001 it was thought to be found only in dry zone and I was surprised to see it in the Talangama Wetland. This shows how little was known about Sri Lankan dragonflies until very recently.

Foggy-winged Twister ■ *Tholymis tillarga*

DESCRIPTION Male has contrasting patch of bluish-grey against black patch on hindwings. Even in low light, this shows up. When hunting, flies low over water inscribing figures of eight. Does not perch on top of vegetation but hangs down from it. At times it may hang on a stem underneath a leaf. Most active at dusk and in some ponds only seen after dusk (however, I have also seen several actively hunting over water during the hottest part of the afternoon). Well known for being attracted to lights in houses. Juvenile male has brownish abdomen that turns red with maturity. **SEXES** Female lacks pale patch on hindwing and is olive-brown on body. End of abdomen is blunt and does not taper to a point, as in male. Terminal appendages spread out and point outwards. **HABITAT** Occurs around ponds, ditches and similar places with still or slow-moving water. Most likely to be seen near waterbodies fringed with thick vegetation. **DISTRIBUTION** Found from lowlands to submontane areas.

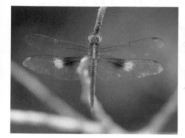

Male

Female

Dingy Duskflyer ■ *Zyxomma petiolatum*

DESCRIPTION Green eyes, blackish thorax and thin blackish abdomen. Very unusual profile with relatively enormous first three abdominal segments, then narrowing down to pencil-thin abdomen. From both above and side, profile is distinctive. Comes out at dusk and is attracted to house lights. Rapid flight, sometimes circling around a pond, low over water. 'Skid-freeze flight', which has a clockwork jerkiness: 'skids' across the air a few inches above water, then 'freezes' in place as it hovers, and skids off unpredictably to freeze again. This unpredictable motion must make it difficult for predators like amphibians and fish to catch it. **SEXES** Similar, but female browner than male, which is almost black. **HABITAT** Seen over ponds and lakes fringed with deeply shaded vegetation. **DISTRIBUTION** Found throughout lowlands. Probably overlooked due to crepuscular habits.

Male

Elusive Adjutant ▪ *Aethriamanta brevipennis brevipennis*

DESCRIPTION Superficially similar to Pink and Spine-tufted Skimmers (see pp. 150 and 148), but considerably smaller. 'Red knees' are diagnostic in male Elusive Adjutant. It is quite an unmistakable dragonfly in terms of its jizz. It has a short, pugnacious look about it because of its build. In male crimson abdomen is like a red beacon, and none of the other red dragonflies seem to have quite the same lustre. Red abdomen is even more conspicuous because head and thorax look a strongly contrasting dark black. In fact, on closer inspection it looks more brownish-black. 'Red knee' on rearmost pair of legs is diagnostic. Male's eyes brownish with tinge of violet at top. In flight it is so swift that it is hard to keep track of – at times I have found it difficult even to make out the colours as it is reduced to a blur. Both sexes have coloured wing-bases, more extensive in hindwings. Male has indistinct black markings on dorsal surface of abdomen. **SEXES** Female has the same build as male, but has a yellowish abdomen richly patterned in black. Female's eyes are brown on top and greenish in lower half, marked with black lines. 'Red knees' are yellow in female. Her hindwings are strongly marked in amber on wing-bases, and to a lesser extent in forewings. Hindwing-bases also have some of cells coloured black. **Behaviour note** Both male and female appear to be faithful to a favourite perch. I have seen what may be the same female at the same perch for more than a week or possibly even a fortnight. Males perch low, beside a vegetated path, exposed to sunlight. Females prefer more shaded perches. **HABITAT** Favours ponds, ditches, lakes and similar places, fringed with vegetation. Tolerant of polluted water. Males and females like to perch away from water. **DISTRIBUTION** Found in lowlands.

Rediscovery of Elusive Adjutant There is an interesting story behind this dragonfly. There had been one record of this species in Sri Lanka and no records after that for over a hundred years. Matjaž Bedjanič, the leading authority on Sri Lankan dragonflies, was contemplating removing it from the Sri Lankan list, thinking that the first record had been a mistake. In 2002 I photographed a dragonfly at Hunas Falls Hotel and asked him to identify the species – it was the Elusive Adjutant. Since then it has turned up all over the lowlands, including in polluted canals in Colombo. As is often the case in natural history, a paucity of records is often a reflection of the paucity of observers.

Female

Male

Scarlet Basker ■ *Urothemis signata signata*

DESCRIPTION Male identified by all-red eyes and black wing-bases. Male has orange-brown pterostigma. Costal edge of pterostigma is black, but pterostigma is never completely black as in male and female Crimson Dropwings (see p. 158). Male Crimson Dropwing shows pronounced red on costal vein and some other veins on leading edge of wing. Spine-legged Redbolt (see p. 156) has a similar pterostigma pattern. However, dark bases to wings are not as dark as in Scarlet Basker, where they are black (enclosing red veins) surrounded by honey-brown. Male Oriental Scarlet (see p. 153) has a similar pterostigma colouration and leading edge veins, but can be separated by amber wing-bases not having black. Furthermore, Oriental Scarlet has thin line of black along dorsal surface of abdomen. Male has black spots on dorsal surface at end of red abdomen. **SEXES** Female yellow overall with thick dark rings on abdomen. Large black patch on hindwings as in male, surrounded by yellow and crisscrossed with yellow veins. Light blue at bottom of eyes. **Behaviour note** Once at Talangama I saw two males perched surprisingly close to each other. A female arrived at the water's edge. One of the males flew up and they mated in the tandem position. They hovered intermittently in mid-air, staying long enough for me to focus on them. The copulation was completed entirely in mid-air. It may have lasted half a minute or a minute at the most, and the female began to oviposit. The male hovered close by, engaging in mate guarding to ensure that no other male would render its sperm void. **HABITAT** Favours ponds, lakes and similar areas, with reeds and bushes to provide perches. **DISTRIBUTION** Found from lowlands to mid-hills. Common.

Female   Male

Sri Lanka Cascader ■ *Zygonyx iris ceylonicum* ℮

DESCRIPTION Black overall, with a reflective body surface – although it is black in colour, sometimes it can catch the light and show a silvery glint. Top of abdomen has

Male

a broken yellow line. I often find this species near forested streams in Sinharaja. It sometimes chooses to fly high and perch high, and has a habit of resting by clinging onto vegetation and hanging down. **SEXES** Female similar to male, but a little bigger. Slightly wider abdomen with yellow lateral stripes on thorax and dorsum of abdomen, and yellow on sides of abdominal segments 2 and 4. **HABITAT** Favours streams in good-quality forests. Quite easy to see in artificial pond near ticket office in Sinharaja. **DISTRIBUTION** Found in wet zone up to lower hills. **STATUS** Least Concern.

Female

▪ CHECKLIST OF THE DRAGONFLIES OF SRI LANKA ▪

The checklist below covers the species known at the time of going to print, although it is highly likely that other endemics await discovery. Since the first edition of the book, 8 new species have been described and are included in the revised checklist below. The older taxonomy at the family level has been retained.

Species in **bold** are covered in this book. Species that are endemic are followed by an 'E'. Species that have an endemic subspecies in Sri Lanka are followed by an asterisk*.

Suborder Zygoptera (Damselflies)

Calopterygidae (Jewelwings)
1   Oriental Green-wing *Neurobasis chinensis chinensis*
2   Black-tipped Flashwing *Vestalis nigrescens* E

Chlorocyphidae (Jewels)
3   **Adam's Gem *Libellago adami* E**
4   Ebony Gem *Libellago corbeti* E
5   **Ultima Gem *Libellago finalis* E**
6   **Green's Gem *Libellago greeni* E**

Euphaeidae (Gossamerwings)
7   Shining Gossamerwing *Euphaea splendens* E

Lestidae (Spreadwings)
8   Scalloped Spreadwing *Lestes praemorsus decipiens*
9   White-tipped Spreadwing *Lestes elatus*
10 Malabar Spreadwing *Lestes malabaricus*
11 Sri Lanka Emerald Spreadwing *Sinhalestes orientalis* E
12 Closed-wing Reedling *Indolestes divisus* E
13 Mountain Reedling *Indolestes gracilis gracilis**

Coenagrionidae (Pond Damselflies or Bluets)
14 White-backed Wisp *Agriocnemis femina*
15 Wandering Wisp *Agriocnemis pygmaea*
16 Sri Lanka Midget *Mortonagrion ceylonicum* E
17 Marsh Dancer *Onychargia atrocyana*
18 Malay Lilysquatter *Paracercion malayanum*
19 Little Blue *Amphiallagma parvum*
20 Asian Slim *Aciagrion occidentale*
21 Dawn Bluetail *Ischnura aurora rubilio*
22 Common Bluetail *Ischnura senegalensis*
23 Painted Waxtail *Ceriagrion cerinorubellum*
24 Yellow Waxtail *Ceriagrion coromandelianum*
25 Malabar Sprite *Pseudagrion malabaricum*
26 Blue Sprite *Pseudagrion microcephalum*
27 Azure Sprite *Pseudagrion decorum*
28 **Sri Lanka Orange-faced Sprite** *Pseudagrion rubriceps ceylonicum**
29 Lieftinck's Sprite *Archibasis lieftincki* E
30 Hanwella Sprite *Archibasis oscillans hanwellanensis**

Platycnemididae (Featherlegs)
31 Yellow Featherleg *Copera marginipes*

Platystictidae (Forest Damselflies)
32 Adam's Shadowdamsel *Ceylonosticta adami* E
33 Austin's Shadowdamsel *Ceylonosticta austeni* E
34 **Bine's Shadowdamsel *Ceylonosticta bine* E**
35 **Brinck's Shadowdamsel *Ceylonosticta brincki* E**
36 Nobel Shadowdamsel *Ceylonosticta digna* E
37 Merry Shadowdamsel *Ceylonosticta hilaris* E
38 **Drooping Shadowdamsel** *Ceylonosticta lankanensis* E
39 Mojca's Shadowdamsel *Ceylonosticta mojca* E
40 Dark Knob-tipped Shadowdamselfly *Ceylonosticta montana* E
41 Nietner's Shadowdamsel *Ceylonosticta nietneri* E
42 Ana Mia's Shadowdamsel *Ceylonosticta anamia* E
43 Bordered Knob-tipped Shadowdamsel *Ceylonosticta submontana* E
44 Blue-shouldered Cornuted Shadowdamsel *Ceylonosticta subtropica* E
45 Dark-shouldered Cornuted Shadowdamsel *Ceylonosticta tropica* E
46 Walli's Shadowdamsel *Ceylonosticta walli* E
47 Alwis Shadowdamsel *Ceylonosticta alwisi* E
48 Hill Shadowdamsel *Ceylonosticta inferioreducta* E
49 Enchanting Shadowdamsel *Ceylonosticta mirifica* E
50 Nancy's Shadowdamsel *Ceylonosticta nancyae* E
51 Rupasinghe's Shadowdamsel *Ceylonosticta rupasinghe* E
52 Pretty Shadowdamsel *Ceylonosticta venusta* E
53 **Dark Forestdamsel *Platysticta apicalis* E**
54 Blurry Forestdamsel *Platysticta maculata* E
55 *Secret Forestdamsel Platysticta secreta E*
56 Serendib Forestdamsel *Platysticta serendibica* E

■ CHECKLIST OF THE DRAGONFLIES OF SRI LANKA ■

Protoneuridae (Threadtails)

57 Two-spotted Threadtail *Elattoneura oculata* E
58 Jungle Threadtail *Elattoneura caesia* E
59 Dark-glittering Threadtail *Elattoneura centralis* E
60 Smoky-winged Threadtail *Elattoneura leucostigma* E
61 Red-striped Threadtail *Elattoneura tenax* E
62 Stripe-headed Threadtail *Prodasineura sita* E

Suborder Anisoptera (Dragonflies)

Gomphidae (Clubtails)

63 Sri Lanka Clubtail *Anisogomphus ceylonicus* E
64 Sinuate Clubtail *Burmagomphus pyramidalis sinuatus**
65 Transvestite Clubtail *Cyclogomphus gynostylus* E
66 Sri Lanka Sabretail *Megalogomphus ceylonicus* E
67 Brook Hooktail *Paragomphus henryi* E
68 Lowland Hooktail *Paragomphus campestris* E
69 Lyrate Grappletail *Heliogomphus lyratus* E
70 Nietner's Grappletail *Heliogomphus nietneri* E
71 Wall's Grappletail *Heliogomphus walli* E
72 Sri Lanka Forktail *Macrogomphus lankanensis* E
73 Keiser's Forktail *Macrogomphus annulatus keiseri**
74 Wijaya's Scissortail *Microgomphus wijaya* E
75 Rivulet Tiger *Gomphidia pearsoni* E
76 Rapacious Flangetail *Ictinogomphus rapax*

Aeshnidae (Hawkers)

77 Pale-spotted Emperor *Anax guttatus*
78 Fiery Emperor *Anax immaculifrons*
79 Elephant Emperor *Anax indicus*
80 Black Emperor *Anax tristis*
81 Vagrant Emperor *Anax ephippiger*
82 Indian Duskhawker *Gynacantha dravida*
83 Millard's Duskhawker *Gynacantha millardi*
84 Dark Hawker *Anaciaeschna donaldi*

Corduliidae (Emeralds)

85 Blue-eyed Pondcruiser
    *Epophthalmia vittata cyanocephala**
86 Flint's Cruiser *Macromia flinti* E
87 Sri Lanka Cruiser *Macromia zeylanica* E
88 Forest Shadow-emerald
    *Macromidia donaldi pethiyagodai**

Libellulidae (Chasers)

89 Fruhstorfer's Junglewatcher
    Hylaeothemis fruhstorferi E
90 Yerbury's Elf *Tetrathemis yerburii* E
91 Sombre Lieutenant *Brachydiplax sobrina*

92 Pale-faced Forestskimmer *Cratilla lineata calverti*
93 Pruinosed Bloodtail *Lathrecista asiatica asiatica*
94 Vermilion Forester *Lyriothemis defonsekai* E
95 Spine-tufted Skimmer *Orthetrum chrysis*
96 Asian Skimmer *Orthetrum glaucum*
97 Marsh Skimmer *Orthetrum luzonicum*
98 Pink Skimmer *Orthetrum pruinosum neglectum*
99 Green Skimmer *Orthetrum sabina sabina*
100 Triangle Skimmer *Orthetrum triangulare triangulare*
101 Blue Pursuer *Potamarcha congener*
102 Asian Pintail *Acisoma panorpoides*
103 Orange-winged Groundling *Brachythemis contaminata*
104 Indian Rockdweller *Bradinopyga geminata*
105 Oriental Scarlet *Crocothemis servilia servilia*
106 Black-tipped Percher *Diplacodes nebulosa*
107 Blue Percher *Diplacodes trivialis*
108 Light-tipped Demon *Indothemis carnatica*
109 Restless Demon *Indothemis limbata sita*
110 Paddyfield Parasol *Neurothemis intermedia intermedia*
111 Pied Parasol *Neurothemis tullia tullia*
112 Spine-legged Redbolt *Rhodothemis rufa*
113 Red-veined Darter *Sympetrum fonscolombii*
114 Crimson Dropwing *Trithemis aurora*
115 Indigo Dropwing *Trithemis festiva*
116 Dancing Dropwing *Trithemis pallidinervis*
117 Aggressive Riverhawk *Onychothemis testacea ceylanica*
118 Sapphire Flutterer *Rhyothemis triangularis*
119 Variegated Flutterer *Rhyothemis variegata variegata*
120 Amber-winged Glider *Hydrobasileus croceus*
121 Wandering Glider *Pantala flavescens*
122 Burmeister's Glider *Tramea basilaris burmeisteri*
123 Sociable Glider *Tramea limbata*
124 Foggy-winged Twister *Tholymis tillarga*
125 Dingy Duskflyer *Zyxomma petiolatum*
126 Elusive Adjutant
    Aethriamanta brevipennis brevipennis
127 Coastal Pennant *Macrodiplax cora*
128 Scarlet Basker *Urothemis signata signata*
129 Sri Lanka Cascader *Zygonyx iris ceylonicum**

Literature consulted

In preparing descriptions of the species, I supplemented my observations in the field by consulting a number of books. The books I found most useful were Bernard d'Abrera's *The Butterflies of Ceylon*, John and Judy Banks' *A Selection of the Butterflies of Ceylon*, L. G. O. Woodhouse's *The Butterflies of Ceylon*, Corbet and Pendlebury's *The Butterflies of the Malayan Peninsula*, Krushnamegh Kunte's *Butterflies of Peninsular India* and Meena Haribal's *The Butterflies of Sikkim Himalaya and their Natural History*.

I would also like to thank Nancy and George van der Poorten, who helped me to identify some of the species I had photographed while working on the first edition of an eight-page booklet on butterflies (2003). Subsequently, in the early years Kithisiri Gunawardana also helped me with the identification of 'confusion species'. Rajika Gamage's *An Illustrated Guide to the Butterflies of Sri Lanka* is useful for its illustrations of food plants. Papers by George and Nancy van der Poorten on the larval food plants were a key up-to-date source of reference. They are listed in the references and cover several species in different families.

The key references for wingspans were L. G. O. Woodhouse's *The Butterfly Fauna of Ceylon* and Isaac Kehimkar's *The Book of Indian Butterflies*. The latter, which is an outstanding piece of work for its coverage, has been very helpful and was kindly gifted to me by Atul Jain, a visitor to Sri Lanka from India. The identification of skippers was helped by this and other books, and the excellent photographs in Laurence G. Kirton's *Butterflies of Peninsular Malaysia, Thailand and Singapore*, from which the forewing diagram (p.11 top left) is taken, and *A Pocket Guide to the Butterflies of Sri Lanka* by Jayasinghe et al. The keys in *The Butterflies of Hong Kong* by Bascombe et al were very helpful.

A number of other books were also consulted and provided additional information. These include *Butterflies of India* by Gay et al, *Butterflies of Ceylon* by Ormiston, *Malaysian Butterflies – An Introduction* by Yong Hoi-Sen, *The Butterflies of Thailand* by Pisuth Ek-Amunay, and two books by Torben Larsen, *The Butterflies of Kenya* and *The Butterflies of Saudi Arabia*.

Butterflies – books

Banks, J. & Banks, J. (1985, several reprints). *A Selection of the Butterflies of Sri Lanka*. Lake House Investments: Colombo.

Bascombe, M. J., Johnston, G. & Bascombe, F. S. (1999). *The Butterflies of Hong Kong*. Academic Press: London.

D'Abrera, B. (1998). *The Butterflies of Ceylon*. Wildlife Heritage Trust: Colombo.

De Silva Wijeyeratne, G. (2006). *Butterflies of Sri Lanka and Southern India*. Jetwing Eco Holidays: Colombo.

Ek-Amnuay, P. (2012). *Butterflies of Thailand*. 2nd edn, English. Amarin Printing and Publishing: Bangkok.

Gamage, R. (2007). *An Illustrated Guide to the Butterflies of Sri Lanka*. Tharanjee Prints.

Haribal, M. (1992). *The Butterflies of Sikkim Himalaya and their Natural History*. Natraj Publishers, Dehra Dun.

Jayasinghe, H., De Alwis, C. & Rajapakshe, S. (2013). *A Pocket Guide to the Butterflies of Sri Lanka*. Chamitha De Alwis: Sri Lanka.

Kehimkar, I. (2008). *The Book of Indian Butterflies*. Bombay Natural History Society and

Oxford University Press: India.

Kirton, L. G. (2014). *Butterflies of Peninsular Malaysia, Singapore and Thailand*. John Beaufoy Publishing: UK.

Kunte, K. (2000). *India – A Lifescape: Butterflies of Peninsular India*. Universities Press (India) Limited.

Larsen, T. (1984). *Butterflies of Saudi Arabia and its Neighbours*. Stacey International: London.

Miththapala, S. (2006). *Butterflies of Sri Lanka for Children*. Text by Sriyanie Miththapala, photographs by Gehan de Silva Wijeyeratne.

Ormiston, W. (1924). *The Butterflies of Ceylon*. H. W. Cave & Co, Colombo.

Owen, D. F. (1971). *Tropical Butterflies. The Ecology and Behaviour of Butterflies in the Tropics with Special Reference to African Species*. Clarendon Press: Oxford.

Watson, E. Y. (1891). *Hesperidae Inidcae: Descriptions of the Hesperidae of India, Burma and Ceylon*. Vest & Company: Madras.

Woodhouse, L. G. O. (1942). *The Butterfly Fauna of Ceylon*. 1st edn. *Ceylon Journal of Science*. Ceylon Government Press.

Butterflies – other publications

De Silva Wijeyeratne, G. (2010). *Butterflies of Sri Lanka*. Sri Lanka Tourism Promotion Bureau, Colombo. A photographic poster illustrating 132 species.

Eliot, J. N. (1973). 'The higher classification of the Lycaenidae (Lepidoptera): a tentative arrangement', *Bulletin of the British Museum (Natural History) Entomology*, vol. 28, no. 6, London, pp. 373–505.

Manders, N. (1904). 'The butterflies of Ceylon', *Journal of the Bombay Natural History Society*, 16(1): 76–85.

Van der Poorten, G. & Van der Poorten, N. (2011). 'New and revised descriptions of the immature stages of some butterflies in Sri Lanka and their larval food plants (Lepidoptera: Nymphalidae). Part 1: sub-family Dananinae', *The Journal of Research on the Lepidoptera*, vol. 44: 1–16.

Van der Poorten, G., & Van der Poorten, N. (2011). 'New and revised descriptions of the immature stages of some butterflies in Sri Lanka and their larval food plants (Lepidoptera: Papilionidae)', *The Journal of Research on the Lepidoptera*, vol. 44: 111–27.

Van der Poorten, G. (2012). 'The taxonomy and conservation status of the butterflies of Sri Lanka', in *The National Red List 2012 of Sri Lanka; Conservation Status of the Fauna and Flora*. Weerakoon, D. K. & Wijesundara, S. (eds), Ministry of Environment, Colombo. pp. 26–41.

Van der Poorten, G., & Van der Poorten, N. (2012). 'New and revised descriptions of the immature stages of some butterflies in Sri Lanka and their larval food plants (Lepidoptera: Nymphalidae). Part 2: subfamily Satyrinae', *Tropical Lepidoptera* Research, 22(2): 80–92.

Van der Poorten, G., & Van der Poorten, N. (2013). 'New and revised descriptions of the immature stages of some butterflies in Sri Lanka and their larval food plants (Lepidoptera: Lycaenidae). Part 1: Polyommatinae and Theclinae (in part)', *The Journal of Research on the Lepidoptera*, vol. 46: 25–49.

Van der Poorten G. & Van der Poorten, N, (2013). 'New and revised descriptions of the immature stages of some butterflies in Sri Lanka and their larval food plants

(Lepidoptera: Pieridae). Part 1: subfamilies Pierini (in part) and Coliadinae', *Tropical Lepidoptera Research*, 23(1): 22–31.

Van der Poorten, G., & Van der Poorten, N. (2014). 'New and revised descriptions of the immature stages of some butterflies in Sri Lanka and their larval food plants (Lepidoptera: Pieridae). Part 2. subfamily Pierinae (in part)', *Tijdschrift voor Entomologie*, 157: 1–25.

Van der Poorten, G.M. & Van der Poorten, N. E. (2016). *The Butterfly Fauna of Sri Lanka*. Lepodon Books. p. 418.

Dragonflies

Bedjanič, M., Conniff, K., Van der Poorten, N. & Šalamun, A. (2014). *Dragonfly Fauna of Sri Lanka: Distribution and Biology, with Threat Status of its Endemics*. Pensoft, Sofia. Essential reference for anyone with a serious interest in Sri Lankan dragonflies. Pdf available online, or print copy can be purchased from Pensoft.

Bedjanič, M., Conniff, K. and De Silva Wijeyeratne, G. (2007). *Dragonflies of Sri Lanka*. Jetwing Eco Holidays: Colombo.

Bun, T. H., Keng, W. L. & Hämäläinen, M. (2010). A *Photographic Fuide to the Dragonflies of Singapore*. The Raffles Museum of Biodiversity Research.

Corbett, P. S. (1999). *Dragonflies: Behaviour and Ecology of Odonata*. Harley Books (B. H. & A. Harley Ltd), England.

De Fonseka, T. (2000). *The Dragonflies of Sri Lanka*. Wildlife Heritage Trust: Colombo.

De Silva Wijeyeratne, G. (2009). 'The dragons of Lanka', *Hi Magazine*, August 2009, series 7, vol. 3, pp. 182–3.

Dijkstra, K.-D. B. & Lewington, R. (2006, reprinted 2010). *Field Guide to the Dragonflies of Britain and Europe*. British Wildlife Publishing: UK.

Dijkstra, K.-D. B., Bechly, G., Bybee, S. M., Dow, R. A., Dumont, H. J., Fleck, G., Garrison, R. W., Hämäläinen, M., Kalkman, V. J., Karube, H., May, M. L., Orr, A. G., Paulson D., Rehn, A. C., Theischinger, G., Trueman, J. W. H., Van Tol, J., Von Ellenrieder, N. & Ware, J. (2013). 'The classification and diversity of dragonflies and damselflies (Odonata)', in Zhang, Z.-Q. (ed.), 'Animal biodiversity: an outline of higher level classification and survey of taxonomic richness (addenda 2013)', *Zootaxa* 3703: 36–45.

Dijkstra, K.-D. B., Kalkman, V. J., Dow, R. A., Stokvis, F. R. & Van Tol, J. (2014). 'Redefining the damselfly families: a comprehensive molecular phylogeny of Zygoptera (Odonata)'. *Systematic Entomology*, 39, pp. 68–96.

Dijkstra, K.-B. & Clausnitzer, V. (2014). 'The dragonflies and damselflies of eastern Africa', *Studies in Afrotropical Zoology*, vol. 298. Musee Royal De L'Afrique Centrale.

Suhling, F., Sahlén, G., Gorb, S., Kalkman, V. J., Dijkstra, K-D. B. & Van Tol, J. (2015). 'Order Odonata', in Thorp, J. & Rogers, D. C. (eds), *Ecology and General Biology: Thorp and Covich's Freshwater Invertebrates*, Academic Press, pp. 893–932.

GENERAL ACKNOWLEDGEMENTS

Many people have over the years helped me in one way or another to become better acquainted with the natural history of Sri Lanka. My field work has also been supported by several tourism companies as well as state agencies and their staff both upon my return to Sri Lanka between December 1999 and May 2010, and during subsequent visits after I returned to work in the UK. To all of them, I am grateful.

My Uncle Dodwell de Silva took me on Leopard safaris from a very young age and got me interested in birds and wildlife. My late Aunt Vijitha de Silva and my sister Manouri got me my first cameras. My late parents Lakshmi and Dalton provided a lot of encouragement and financial support for expensive books and equipment, and my sisters Indira, Manouri, Janani, Rukshan, Dileeni and Yasmin, and brother Suraj, also encouraged my pursuit of natural history. In the UK my sister Indira and her family always provided a home when I was bridging islands. Dushy and Marnie Ranetunge also helped me greatly on my return to the UK in June 2010. Since then I have received support and inspiration from many people. I thank all of them.

My one time neighbour Azly Nazeem, a group of then schoolboys including Jeevan William, Senaka Jayasuriya and Lester Perera, and my former scout master Mr Lokanathan were a key influence in my school days. This book is a continuation in the spirit of the simple pictorial guide booklets and posters published several years ago by Jetwing Eco Holidays. The business and design skills of Chandrika Maelge for these and other publications helped to create livelihoods in wildlife tourism and brand Sri Lanka as a wildlife super-rich destination.

A book such as this inevitably draws on various books. I am grateful to authors who have worked on butterflies and dragonflies all over the world, who have made my task easier. My wife Nirma and my two daughters Maya and Amali, at times with help from parents Roland and Neela Silva, take care of many things, allowing me time to work on books.

Tara Wikramanayake helped by copy editing the preliminary drafts of this book, diligently alerting me early on to sections that had gaps or places where the text lacked clarity. Krystyna Mayer once again provided a thorough edit. John Beaufoy and Rosemary Wilkinson were supportive and persuaded me that in spite of a busy day job, I could find the time to write this book and take photographs for it.

Specific acknowledgements, butterflies
Shayani Weerasinghe and Nirusha Ranjitkumar assisted with compiling information on wingspans. Rohan Pethiyagoda made available an unpublished checklist on the flora, which has been useful. Priyantha Wijesinghe shared his unpublished checklists of butterflies, which initially helped me to catalogue my images systematically, and I have used this older taxonomic arrangement. Sriyan de Silva Wijeyeratne kindly invited me to write an article for his book with a view to adapting it for this book. Bernard d'Abrera's *The Butterflies of Ceylon* was in my field bag; it was a useful source of reference for many years.

Specific acknowledgements, dragonflies
Karen Conniff and Matjaž Bedjanič have assisted me in the identification of species for over a decade. I am very grateful to them. Matjaž Bedjanič kindly reviewed the draft of the dragonfly section in this book. Any errors that remain are mine.

Numerous visits to Talangama Wetland in the company of other photographers to obtain images of butterflies and dragonflies have benefited from the hospitality of Sherine and Preethi Perera, Hema de Silva and his staff at Villa Talangama, and Karen Conniff and David Molden. Sherine also took care of the One Acre reserve in which many of the photographs were taken. Tilak Conrad and Keith Blom and their staff at Copyline worked hard on the photographic guides to butterflies and dragonflies published by Jetwing Eco Holidays, which laid the foundations for this title.

Corrections to 1st Edition

My thanks to Matjaž Bedjanič for having drawn my attention to the following mistakes in the first edition.

p. 154 Black-tipped Percher *Diplacodes nebulosa* female was a Blue Percher *D. trivialis* female.

p. 155 Light-tipped Demon *Indothemis carnatica* female was an Orange-winged Groundling *Brachythemis contaminata* female

p. 159 Indigo Dropwing *Trithemis festiva* female was a Crimson Dropwing *T. aurora* female.

Request for Images

If you have images of a species (or a male/female/juvenile) not shown in this book and would like to share these images for a future expanded edition, please contact the author on gehan.desilva.w@gmail.com.

TOUR OPERATORS

A non-exhaustive list of just over 20 companies that the author is acquainted with is given below.

A Baur & Co. (Travels), www.baurs.com
Arunya Vacations, www.aarunyavacations.com
Adventure Birding, www.adventurebirding.lk
Aitken Spence Travels, www.aitkenspencetravels.com
& Beyond, www.andbeyond.com
Birding Sri Lanka.com, www.birdingsrilanka.com
Bird and Wildlife Team, www.birdandwildlifeteam.com
Birdwing Nature Holidays, www.birdwingnature.com
Eco Team (Mahoora Tented Safaris), www.srilankaecotourism.com
Hemtours (Diethelm Travel Sri Lanka), www.hemtours.com
High Elms Travel, www.highelmstravel.com
Jetwing Eco Holidays, www.jetwingeco.com
Lanka Sportreizen, www.lsr-srilanka.com
Little Adventures, www.littleadventuressrilanka.com
Natural World Explorer, www.naturalworldexplorer.com
Nature Trails, www.naturetrails.lk
Quickshaws Tours, www.quickshaws.com
Red Dot, www.reddottours.com
Sri Lanka in Style, www.srilankainstyle.com
Walkers Tours, www.walkerstours.com
Walk with Jith, www.walkwithjith.com

▪ INDEX ▪